Tireless focus, for a moment of strategic opportunity

HUAWEI

Focus · Persevere · Breakthrough

Wagenia man fishing in the Congo River

Praise for the Book

"One of the attributes of Huawei is that the company has gone global, reaching a relevant position in Europe and addressing the [United States]. The US government, however, is slowing down the process of Huawei entering the [US] market on suggestions that Huawei could get access to critical US government information. No question Huawei is the result of the entrepreneurial effort of Ren Zhengfei. The book allows [one] to get some understanding of the entrepreneur, his philosophy, the culture he wants to establish in Huawei, and the process to do it.

Few of the leading Chinese companies have taken relevant positions internationally. For a high-technology company, addressing the international markets is an important challenge. But Huawei has managed to have thousands of employees outside China and satisfy important customers in many countries. The book helps in understanding the approaches followed by Huawei. Mr. Ren Zhengfei has been capable of pioneering the development of technology while growing in the world. Obviously this has required a strict control of information, which attaches a flavor of secrecy and even mystery to the behavior of Mr. Ren Zhengfei. He has maintained the company as private and his management style has been extremely unconventional.

The book by Tian Tao and Wu Chunbo is an extraordinary research effort to analyze and describe a very complex reality and a very special entrepreneur. It is a very valuable work to understand entrepreneurship in our complex world."

— Dr. Pedro Nueno,
President, Chengwei Ventures Chair Professor of
Entrepreneurship, CEIBS, China

"Huawei is a prominent company, among the most successful and most internationalized in China. This book offers insights to Western readers, allowing them to truly understand Huawei, its management philosophy, and culture, and the special leadership approaches of Ren Zhengfei."

— John A. Quelch,
Charles Edward Wilson Professor of Business Administration,
Harvard Business School

"Over the course of the last quarter century, Huawei has become one of the most important telecommunications companies in the world and, arguably, one of the most powerful Chinese companies in the global economy. Yet, there is much that is not known about Huawei and much that is misunderstood. As the foremost authority on Huawei in the world, Tian Tao brings deep knowledge and unprecedented access to help pull back the veil on one of the world's most enigmatic companies. This book will be essential reading for politicians, pundits, and anyone interested in gaining a deeper understanding of this important company."

— Dr. Doug Guthrie,
Professor and former Dean at the
George Washington School of Business
Author of *Dragon in a Three-piece Suit: The Emergence of
Capitalism in China*

"As a long-time observer and friend of China, I have watched with admiration the global growth and innovative strategy of Huawei. This book will give readers a special insight into Huawei's capacity and entrepreneurial drive under the dedicated leadership of Mr. Ren Zhengfei."

— Kerry Matthew Stokes AC,
Executive Chairman, Seven Group Holdings Limited, Australia

"This is the most complete book about Huawei I have ever read."

— Liu Chuanzhi,
Founder of Lenovo

"Over the past two decades, Huawei has composed an amazing epic in the business world. It started from an emerging market and conquered the global market. What is the code of success? This book offers the most reliable answers."

— Qin Suo,
Mediaperson, former Editor-in-Chief of *CBN Daily*

"This book does not present the business history of Huawei only; it also highlights the evolution of the business philosophy of Ren Zhengfei, the boss of the company. Huawei is Mount Everest in the mind's eye of all Chinese businessmen, and Ren Zhengfei has behaved like a hermit living in a cave. He has not created any precept or claimed any truth. As the closest observers of Ren Zhengfei, Tian Tao and Wu Chunbo are the most likely authors to tell us the precepts and truths of Ren Zhengfei."

— **Niu Wenwen,**
Publisher and Creator, *The Founder Magazine*

"Huawei is a private-owned company. It is highly sensitive to technological changes. It is the real multinational based in China. These facts reflect the alertness, adaptability, and conviction of Ren Zhengfei, the top leader of Huawei. Ren is a unique manager and his speeches have been widely quoted because they offer great philosophical delights. This book is full of such delights, which you cannot miss."

— **Liu Zhouwei,**
Founder, *21st Century Business Herald*

Huawei

Huawei
Leadership, Culture, and Connectivity

TIAN TAO
DAVID DE CREMER
WU CHUNBO

Los Angeles | London | New Delhi
Singapore | Washington DC | Melbourne

First published in 2017 by

SAGE Publications India Pvt Ltd
B1/I-1 Mohan Cooperative Industrial Area
Mathura Road, New Delhi 110 044, India
www.sagepub.in

SAGE Publications Inc
2455 Teller Road
Thousand Oaks, California 91320, USA

SAGE Publications Ltd
1 Oliver's Yard, 55 City Road
London EC1Y 1SP, United Kingdom

SAGE Publications Asia-Pacific Pte Ltd
3 Church Street
#10-04 Samsung Hub
Singapore 049483

Published by Vivek Mehra for SAGE Publications India Pvt Ltd, typeset in 11/14 pt GillSans by Fidus Design Pvt. Ltd., Chandigarh 31D and printed at C&C Offset Printing Co., Ltd., China.

Library of Congress Cataloging-in-Publication Data Available

ISBN: 978-93-860-6205-5 (HB)

SAGE Team: Clare Sun Suqing, Sandhya Gola, and Rajinder Kaur

*Vision, will, and integrity—gifts for young men
and women from Ren Zhengfei*

Contents

Simon Murray

Foreword

This is a story not of a company but of a man. A man possessed of all the qualities that we need in our leaders, but so seldom find— vision, courage, determination, honesty, integrity, tolerance, fortitude, and more, much more. This is a compulsory read for all students of life, not those who want to "be" but those who want to "do."

Ren Zhengfei, having left the army, as a relatively low-ranking officer with an engineering background, started his business with only CNY20,000 in 1987. The mobile telephone was going to enter the market. The big "dinosaurs," such as AT&T, BT, Cable & Wireless, France Telecom, Deutsche Telekom, and all the rest of them, who had been milking their monopoly status for years, were about to be invaded by ants heralding the arrival of the handheld cellular telephone. One of these in China was Huawei, a private start-up with three employees and a wafer-thin balance sheet, looking for space in a competitive field full of the companies from the United States, Europe, and Japan as well as Chinese state-owned companies supported by the central government, both financially and politically.

Today, with 170,000 staff worldwide, Huawei's 80,000 employees are stockholders in the company, with shares given to them by Mr. Ren, who has kept only 1.4 percent for himself. This makes Huawei a 100-percent employee-owned company. It now has revenues of US$40 billion and operates in over 170 countries around the world.

This book is not only about amazing corporate success and the reasons for it, but also about philosophy, wisdom, humility, and a wonderful culture from which many in the Western world could learn. The United States accuses Huawei of "listening" to others through their systems, but Mr. Ren is far too shrewd to do that, for he knows that if he did and was found to be doing so, even once, his business would die. He leaves that to the others!

We know dinosaurs die—Motorola, Lucent, IBM, and so many others. They get distracted, go down the wrong path, or forget that the priority is not the technology but the customers. This book examines some of those stories. It may sound like a heavy subject, but it is not. It is a racy read, full of the excitement of competition, blended with the quiet philosophy and observations of the wonderful Mr. Ren, who makes other business "heroes" look metallic and devoid of warmth, which he has in abundance.

Simon Murray is the chairman of GEMS Ltd and Gulf Keystone Petroleum Ltd. Murray in the past held various high positions on boards, including group managing director of Hutchison Whampoa Ltd, non-executive chairman of Glencore International AG, executive chairman (Asia-Pacific) of the Deutsche Bank Group, and vice-chairman of Essar Energy plc. At age 63, he became the oldest man to reach the South Pole unaided, on a 58-day slog, during which he lost 50 pounds (23 kilograms).

Admiral William (Bill) A. Owens

Foreword

The information and communications technology (ICT) sector is fast-paced and relentless in transforming its prominence in our everyday lives, and it is causing our lives to be more efficient and fulfilling. Since the invention of the telephone, consumers have benefited greatly from the evolving scale of this technology platform. It is natural for us to instantaneously access information and make important decisions on demand. Our reliance on ICT's ubiquity and reliability in this age of the "Internet of Things" has concurrent expectations for speed and content accuracy. Prospects for these elements and the baseline for how we communicate continue to spiral upward. This connectivity is making our world a better place and one of the most significant driving forces behind this swelling technology wave is Huawei.

A book about Huawei is a book about its founder. Ren Zhengfei has made a difference that many will never appreciate in the building of what is likely the most successful private business to come out of post-Mao China. Ren is a humble man with humble beginnings, he is a proud Chinese, a former People's Liberation Army (PLA) soldier, but unlike how the Western press would portray Ren, he is not a career military officer and was not a high-ranking "general" as some in the West have wrongly represented him as.

I have known Ren Zhengfei for over 10 years and have followed Huawei with a sense of curiosity and respect for their corporate culture, their core values, and their innovation. I competed head on with Huawei while I was a CEO at Nortel Networks and realized early on that Huawei's motivation to succeed in the telecom space was resolute, with very specific intentions. Ren led Huawei with a master's blend of hurried patience, innovation, and focus, which moved the company forward in large, steadfast steps. Now at the forefront of the telecom industry on a global scale, Huawei remains fiercely committed to innovation and operational excellence.

My relationship with Ren began in a set of meetings the two of us had concerning the potential merger of Huawei and Nortel, a merger that was not to be; yet, my time with him allowed me to get to know him personally, and our friendship was established. I can remember many meetings in mahogany boardrooms and discussions with Ren related to Chinese folklore about "chicken ribs," but what we didn't lose in translation was a wisdom that he brought to the table, a story of dedication, global strategy, and thoughtfulness that caused me to reflect on the certainty that Huawei, under his leadership, would one day be the best in the business.

Ren has many similarities to some other great people I have known. Outside of Ross Perot, Sr.'s office is a stunning picture of an American bald eagle flying solo over a beautiful Pacific Northwest scene with the caption"Eagles Don't Flock". Certainly, Ren has never flocked! This book skillfully points out those elements of folklore and wisdom.

Huawei, the "wolf," has competed fiercely around the world, and alongside Ericsson, has essentially become a formidable market leader in advanced wireless and fiber telecom equipment. This has been skillfully done, through a unique culture, a dedication to the business (only occasionally seen in the West), a strong management efficiency, precise execution, and individual responsibility and accountability.

I believe that Huawei embodies a new management strategy, the one that is based on culture, a culture of patience and humility, a culture based on doing everything possible to delay an "almost certain failure in the future" and a "go-it-on-your-own" approach. This philosophy and the complete reliance on the Huawei employees and employee ownership also make an IPO (initial public offering) difficult, because it would put the destiny of the company in the hands of outside shareholders and an unpredictable marketplace, one that tends to rely on a quarter-to-quarter strategy and results.

"Dedication" is a profound word, and Ren has inspired his people with the spirit of unwavering dedication and commitment to the business and their customers. Often to abstraction, he has focused on the business and caused his people to do the same.

Ren is clearly focused on the future. You can see this from the continued dedication of research and development (R&D) resources to the development of products. He has learned this and other precepts from Western consultants, but I think it is also a part of his natural makeup. He knows this is necessary in order to pursue that dedicated long-term vision of a company that will continue to be, in the future, one of the two or three most significant telecom and enterprise technologies companies in the world.

As a military man, I have known many clever and truly outstanding strategists. I have rarely come across an individual more strategically oriented than Ren. He studies the world, studies the market, talks with many people with knowledge, and then judges where the pitfalls and minefields lie for his business. This has led the company to take great advantage of both the upsides and downsides of this industry. He has forecasted both booms and busts and has been unafraid to take contrarian action to drive his company through them. After 30 years in the US military, I have some idea of grand strategies. Ren has managed a grand strategy that has continued over the decades of his leadership and included all elements of politics, finance, markets, and a familiarity with the black swans stalking the business. This book cleverly lays out those elements of brilliance that he has brought to the marketplace and should be another great reminder to many MBA classes about what real strategy can do.

Huawei's impressive performance as a corporation has led some countries, most notably Great Britain and the United States, to raise concerns over Huawei's alleged capability to collect sensitive information through its network equipment. These contentions have not been proven and are largely wrapped in a political mesh that has excluded Huawei from competing effectively in these markets. Notwithstanding, under Ren's leadership, Huawei has become a global leader in the telecom sector (second only to Sweden's Ericsson). This competitive ranking, while excluded from some of the largest international opportunities, speaks of Huawei's formidable achievement.

I believe, especially, that in the United States there is a growing awareness that in the telecom equipment industry, Huawei must be allowed to enter our country in a meaningful way to allow us to take advantage of the world-changing technology that they have developed and will continue to develop in the future. The United States is no longer leading the world in telecom and enterprise equipment build-outs. It's time to put the history of charges and countercharges surrounding Huawei and the Western governments around the world behind us. Huawei possesses technologies that the United States needs, and the partnership with Huawei could be profound for them and the technology companies in the United States.

In conclusion, the story of Huawei so well told in this book is about an employee-owned company, a humble founder and leader, and a different philosophy, one that we should all consider as we look toward new approaches to our businesses in an unpredictable world.

Bill Owens is the chairman of one of the three largest telecom companies in the United States, and has been a board member of 23 public companies. He was a nuclear submarine officer in the US Navy and retired as a four-star officer, the vice chairman of the Joint Chiefs of Staff in the United States. Since leaving the navy, Owens has served as a CEO in three companies, including a Fortune 500 company.

Mikael Lindstrom

Foreword

After years of phenomenal growth, China has arrived at a point where it is a middle-income economy. Its big challenge is to reform the driving forces of the economy so as to be able to continue the upward journey, rather than becoming stuck in the so-called "middle-income trap." A key to success is developing innovative products and solutions, building global brand recognition and independent marketing structures. Japan and, later, Korea succeeded in doing this. Can China do the same?

We still don't have the answer. However, the story of Huawei indicates that it can be done. Huawei happened in a time and a place that were unique in Chinese history. Founder Ren Zhengfei left his job as an army engineer when the Chinese military–industrial complex was radically downsized. He started a small import business in the commercially most liberal place in China, the special economic zone of Shenzhen, just across the Hong Kong border. He made the most of the opportunity that history provided. He chose an operating model that had never been seen in China, sharing ownership with his employees. Huawei's growth is the result of dedication and pragmatism of Ren and the other early employees.

Reading Tian Tao's account of what has driven Huawei to where it is today provides more insight than anything I have seen before about Huawei. His isn't the traditional type of corporate history that we know well, the ones that start with the family background of the founder, then focus on early challenges, maybe describing some near-death experiences, before moving on to ultimate success. Rather than delving into sales statistics, product development, and so on, Tian Tao focuses on the evolving philosophies and strategies that have guided the company. It is a fascinating inside account that explores the soul of the company.

It is a story told in Chinese way—there are many metaphors—and from Chinese perspective, which here and there will surely surprise a Western reader. But this has the great advantage of helping us understand how China sees the world. The many stories about the Chinese corporate landscape are an additional bonus.

A dominant theme is how quickly the wheel turns in business. Firms grow, may become immensely successful, but most often then flounder and disappear. The telecom sector has had more than its share of such developments. The list of failed information and communications technology (ICT) giants is long. Huawei itself was close to leaving the ICT sector—in the book, I learned that Huawei in 2002 agreed to sell its hardware system to Motorola for US$10 billion, investing the proceeds in real estate. A new Motorola chairman turned down the deal at the very last minute. Imagine how it would have, had it gone through, changed the course of both companies!

Awareness of the continuous cycle of growth and death has very much influenced Ren Zhengfei. A constant question in his mind has been how Huawei could succeed when so many others failed. Humility and curiosity have been guiding sentiments. Or to use a rather Chinese term, "self-criticism." Ren Zhengfei wrote in 2008, "Without self-criticism, we would have confined ourselves to an enclosure and missed out on the new ideas which have proven essential for us to become a world-class corporation. … Our future is dependent on how long we keep the tradition of self-criticism."

One of the answers to the many challenges that have confronted the company has been the willingness to learn from the positive experience of others, not just in the area of technology but also in soft matters. Huawei has made a major and multi-year effort to learn and adopt Western management techniques. It also understands that, as a global company, it needs to integrate into the societies where it is active. A quote from Tao Tian tries to capture this: "Huawei leverages the best from China and the West, and yet it is neither Chinese nor Western."

My personal acquaintance with Huawei is as one of 15–20 foreign senior advisors. We meet Ren Zhengfei and the corporate leadership every year for a polite but frank dialogue of a kind that I doubt takes place anywhere else in Chinese industry. This willingness to listen to and debate with foreigners with widely varying backgrounds speaks tons about the openness that characterizes the company.

Mikael Lindstrom is a senior advisor to Six Year Plan AB and a former Swedish ambassador to China.

No One
Is an
Authority
on the
Huawei Story

The story of Huawei was first published at the end of 2012 to great reception, both at home and abroad. The book was translated into Traditional Chinese, Korean, English, and Japanese, resulting in mixed reviews from tens of thousands of readers across various media, including microblogs, WeChat, books, newspapers, and journals, sparking a new wave of research on Huawei in both academic and business circles.

What I found most interesting was that, outside of Huawei, the book received mostly positive reviews, whereas internally—especially on Huawei's online forum, Xinsheng Community—the majority of comments were critical, sarcastic, and some were even quite cutting. Representative comments include:

1. The book is far too kind in its depiction of Huawei: "It reads suspiciously of company propaganda."

2. Some facts and points of view are not accurate, and might "have an element of subjectivity or imagination."

3. The book's title doesn't reflect the content: "The title feels like clickbait."[1]

4. "The book is a very superficial look at Huawei—it errs on the shallow side."

5. The book is not well-structured and some content overlaps.

---

[1] This comment refers to the name of the original Chinese edition, which translates roughly as *Will Huawei Be the Next to Fall?*

To be honest, after reading such overwhelmingly negative reviews on the Xinsheng Community forum, I felt a surge of emotional resistance. This book was the result of over a decade's worth of close observation, poring over countless public and semi-public company documents that amounted to millions of words. It took six straight years of thinking and writing to finish. How could it be "superficial," "shallow," or come off as propaganda?

When working on the first edition of this book, from start to finish, I was operating half in secret. Judging from the low profile that Huawei is well known to keep, I was worried that Ren Zhengfei and his senior executives might be against my writing such a book and "persuade" me to kill the idea. In fact, in the past 10 years, quite a few books on Huawei had also been "dealt with" in this way, but the authors refused to comply. If Huawei had indeed tried to make me give up on publishing this book, I'm not sure what I would have done.

As is often the case, though, Huawei managed to surprise me. After the first draft of the book's manuscript was sent to the rotating CEO, Xu Zhijun, and then forwarded on to dozens of mid- to senior-level managers, the feedback to me was: publish or not, it's up to you. I suddenly realized that Huawei, caught up in a storm of globalization, was about to fully open itself up to the outside world. Indeed, over the two years that followed, Huawei completely reversed its position of keeping distance from the media. The rotating CEOs started accepting high-profile interviews from domestic and foreign journalists. Later, even the "mysterious and reserved" Ren Zhengfei put himself in the spotlight and had a few candid exchanges with journalists both at home and abroad, culminating in the Davos Forum in early 2015, where he did a live global interview with a well-known BBC reporter. Over the past two years, Huawei has become one of the topmost watched companies of its size among international media.

Of course, Huawei has been quite successful in its "media desensitization" treatment: Its transparency has led people to finally realize that this company from China is worthy of respect. In fact, the company is rather unique in terms of its business and management philosophy, culture and systems, approach to innovation and reform, as well as its evolving strategies, tactics, and leadership. Consequently, Huawei has become a highly cited case for business programs around the world, and in recent times, I've been invited to discuss Huawei at depth with professors from over 10 different business schools.

To my delight, I was given a special privilege after the first edition of the book was published: I received unrestricted access to one-on-one, face-to-face interviews with Huawei's management team. It was a massive undertaking, lasting a total of 19 months and involving people from sales, R&D, finance, legal, auditing, administration, and logistics, across all sorts of departments and product lines. Over 100 managers granted me interviews, including almost every member from Huawei's board of directors and its supervisory board. Each interview took 3–6 hours, the transcriptions alone adding up to nearly 3 million Chinese characters. Having gone through this, I can say with confidence that, among all the non-Huawei people out there, I know Huawei the best. Even among Huawei's employees numbering more than 170,000, I am one of the very few who can perceive Huawei's many different facets from multiple perspectives.

About halfway through the interview process, however, I began to feel uneasy. And after they were all done and over with, as I went through my notes, I had a sinking feeling about the coarse and unfinished nature of the first edition. I could hardly bring myself to go back and read some of its chapters, and I finally came to terms with the criticism and feedback I received from Huawei's people on the Xinsheng Community forum.

In early June of 2014, I began to mull over the idea of chopping up the entire thing and revising its content. My first plan of attack was to straighten out the historical context. In order to get to the heart of certain events, I wanted to cross-reference interviews from at least three different sources. Second, when forming conclusions and opinions, because so much information was suddenly available to me, I tried to be as precise and objective as possible. Of course, after making these revisions, I had to go through and restructure some parts of the book. In total, revised and newly added content accounts for more than two-thirds of this new edition.

Revising the book was no less difficult than writing the first edition. To be sure, the revised version is essentially a new book, with an overarching structure. Amidst the chaos of over 3 million words' worth of stories, opinions, and details, I was often at my wit's end. The restructuring of chapters and sections, the corresponding facts and conclusions—all of these added up to many sleepless nights. I would often get up at 2 AM or 3 AM to write or go over interview extracts. After a few months of this, I lost a ton of weight and my face wouldn't stop twitching.

Before the Chinese New Year in 2015, I finished the first round of revisions and read through the whole book more than five times (going through the details of certain chapters over and over again). At that time, a good friend of mine pointed out that the revised book sort of fizzled out towards the end, so on the third day of the Chinese New Year, I set out to completely rewrite the final two chapters. Afterwards, I sent the yet-again revised version to a dozen or so people (both inside and outside of Huawei) for review, and for another three months, underwent five new rounds of revisions based on their copious feedback.

If I were able to give each book a score, I think the first edition would at most get a 60 out of 100, but the new edition would be a solid 80. You might wonder why I wouldn't give my own book a higher score. And the reason is simple: Huawei has a rich history and culture—far too much to capture in a single book. And moreover, Huawei is constantly changing. When you think you've finally captured the "ultimate truth," if you look at it from another perspective, it's likely that you just captured a fallacy, or that a so-called "truth" was recently discarded in light of evolving views and new models. Essentially, no one can ever claim that they are an "authority" on Huawei. And after interviewing over 100 people, I became even less confident—not only about the facts, per se, but also about some of the conclusions I drew. I can't say for certain that either set represent the full truth.

On the other hand, this gave me a different kind of confidence: Being fully aware of my deficiencies, instead of feeling anxious and uneasy, I would rather open myself up to scrutiny and invite the 170,000-plus Huawei people (including tens of thousands of former Huawei employees) as well as all my friends and all the followers of Huawei out there to come and comment on this revised version of the Huawei story. Have at it. Your criticism and suggestions on every detail and viewpoint in this book will help make the next revision all the more objective and authentic.

Tian Tao

| would like to thank the following Huawei people for making this book possible: Yin Zhifeng, Su Baohua, Cao Yi, Hui Caihong, Shen Shengli, Zuo Fei, Chen Danhua, Zhang Junjuan, Gong Hongbin, and Ye Xiaowen. Over the past two years, they have worked with me to complete 136 interviews with Huawei's management team (including a few individuals who had already left the company). From putting together the list of interviewees, to discussing the questions, arranging the interviews, and structuring the content we gathered, they spent a great amount of time and effort in helping me secure a diverse, rich assortment of information on Huawei. Without the outstanding contribution of these 10 ladies and gentlemen, this book would not have been what it is today.

Of course, I would also like to thank the current and former Huawei people who agreed to be interviewed. Although there are vast differences in their positions, ages, and personalities, they all have one trait in common: To my surprise, they were all quite candid and not at all reserved. This was far more than I had expected, and it was a huge relief. Naturally, this has helped considerably in ensuring the objectivity of the book and has helped set a solid, factual foundation for future research on Huawei.

It is unfortunate that owing to space constraints and the necessity of extracting pertinent content, a large portion of the material from the interviews was not covered in this book. This material will have to wait for the next book, or appear in the results of other researchers who study Huawei.

Thanks also to management scholar Ms. Chen Chunhua and Professor Anthony Tseng Tsai Pen at the National University of Singapore for their

Acknowledgments

guidance throughout the writing process. Without the encouragement of these two fine teachers, it would have been impossible for me to have made such a great leap.

Wu Xiaobo, the dean of the Zhejiang University School of Management, was kind enough to offer his invaluable insight into innovation and transformation, which was a source of deep inspiration for many concepts presented in this book. Mr. Wu is a fine mentor and a good friend, and I would like to thank him very much.

And a special thanks goes to my wife, Ms. Yao Baochen. Her constant encouragement both inspired and compelled me, even when I was too tired to go on. At one point, I had even considered dictating the rest of the book to someone else, but when I lacked the strength to go on, she would help me regain mental fortitude and overcome myself, at last producing the first edition of the Huawei story and the version you see before you today. I offer her my most heartfelt gratitude.

Tian Tao

Business Management Philosophy: The Mysterious Power behind Huawei's Success

Introduction

Time and Destiny: The Delusional Ren Zhengfei

Back in 1987, a middle-aged man named Ren Zhengfei was abandoned by mainstream society. He was 44 years old. This is a significant age, because in Chinese the number four sounds the same as the Chinese character for death—"to be released from the mortal coil," or "to vanish." But double the number four and you've got the number eight, which sounds like the Chinese character for prosperity in the southern dialect. Four times four is 16, which means "the road ahead is smooth."

This age, rich as it is with numerological metaphors concerning the great mysteries of fate, marked the beginning of Ren Zhengfei's entrepreneurial journey. Prior to this, he had already spent three years in the business world, but that experience had ended in failure.

In 1987, this 44-year-old failure founded Huawei Technologies, a private company, and from that point forward the very destinies of Ren Zhengfei and of Huawei itself merged together as one. In the early days of China's great reform, Huawei was dismissed as a privately owned "bastard of a business," widely viewed as an illegitimate pursuit in a time of economic transformation. Ren Zhengfei himself was some sort of delusional Don Quixote, lance in hand, tilting at windmills with unrealistic dreams of grandeur, against all but impossible odds.

In the late 1970s, Deng Xiaoping launched a disruptive economic revolution in China. Within a decade, from 1978 to 1988, the country's entire economy and social structure underwent drastic change. For thousands of years, China had been dominated by politicians and men of letters, and agriculture was the key or even the sole source of economic power. Merchants, as a profession, had no position in the country, and even the most successful merchants such as Lü Buwei (292–235 BC) and Hu Xueyan (1823–1885) survived only as political dependents. Through his reform, Deng Xiaoping helped China break through the edifice of this rigid tradition.

Deng Xiaoping is acclaimed for his political courage and foresight because he was the driving force behind this large-scale commercial revolution that had far-reaching impact in China and abroad. His reform was a ray of hope for merchants who, as a social class, have since gradually secured a more prominent position in the country.

In the 1980s, people from all different backgrounds in China were stirring with enthusiasm and restlessness. There emerged a number of key reformers, the first movers and shakers who "dared to eat crab," as they were described, which at the time meant to try something new in spite of its menacing appearance. Among these were Bu Xinsheng, Ma Shengli, Mou Qizhong, Nian Guangjiu, Zhang Ruimin, and Liu Chuanzhi. China's economic reform threw this group of adventurers and rebels together into a kaleidoscopic whirlwind of the new and unknown.

At this point in time, former military engineer Ren Zhengfei was also drawn into this whirlwind of marginalized businessmen, whether he liked it or not. And as a private business owner, at that.

In its formative days, Huawei started out as telecom equipment trader. In the climate of early reform, this type of trade was often looked down upon in the same way that we might now look down upon door-to-door salesmen. All the same, Ren held high expectations for himself and this little enterprise of his: to become a world-class telecom equipment manufacturer within two decades.

If heroes are born in times of chaos, then who ends up rising as leaders in the turbulent days of reform? Nassir Ghaemi, a professor of psychiatry at Tufts University, says that such leaders are often found among the mentally ill. He posits that outstanding leaders, during critical moments in history, are often afflicted with psychological illness. This is also true of Ren Zhengfei, who suffered from serious depression and anxiety during the darkest days at Huawei. More than 20 years ago, when Ren announced his dream of becoming a world-class company, his small audience was divided; the "believers" were only half-convinced, while others simply thought that their boss was off his rocker.

Of course, in those days there were many delusional visionaries like Ren Zhengfei, as his was a passionate and peculiar generation. Liu Chuanzhi, the founder and chairman of Lenovo, was among the delusional. Then involved in computer parts trade, his "neurosis," so to speak, was to one day challenge IBM. Twenty years later, he did it. Another of the clinically insane was Mou Qizhong, who once declared he would blast apart the Himalayas to channel water from the Brahmaputra River to the Yellow River, and who once traded a train full of Chinese socks for Russian jets. He ended up in prison, but according to recent reports, not even the prison walls or their barbed wire could suppress his passion and flights of fancy.

The telecom industry was no stranger to madmen. Take a look at some of the names they gave their companies. There was Great Dragon, the mythological symbol of China—and "great," no less. Then there was Datang, with "tang" referring to the Tang Dynasty, one of the most powerful and prosperous dynasties that, to some extent, represented China at its best. Then there was Zhongxing Semiconductors (now the ZTE Group)—which literally means "China is reviving." Last but not the least, there was Huawei, meaning "China has a promising future."

The first of these Chinese telecom equipment manufacturers—the four most iconic and successful in China—was founded in 1985 (ZTE). They all started with the odds stacked against them, suffering from a lack of capital, technology, and human resources. To make matters even worse, they faced immensely powerful competition from Western equipment manufacturers, almost all of which had been established for over a century. They were like a handful of ants taking on a herd of elephants—only in their wildest dreams could they have ever emerged victorious.

Now let's take a look at these four middle-aged, imaginative idealists in a world more than 20 years ago: Wu Jiangxing of Great Dragon, Zhou Huan of Datang, Hou Weigui of ZTE, and Ren Zhengfei of Huawei. Over the past two decades, these four men played pivotal roles in the Chinese and global telecom markets, a tragicomic journey full of rampant twists and turns.

All Roads Lead to Rome: The Story of Four Middle-aged Men

Arrogant Multinational Giants

In the 1980s, China was at the forefront of revival. Revitalization was a buzzword in the political and media spheres of a country eager to recover from political tragedy. Revival was the shared aspiration of people at all levels of society.

At that time, economic activity had stagnated for decades and the social system was defunct. Deng Xiaoping's call to reform and open up brought the entire country to attention, liberating its productive forces and, more significantly, liberating its people from rigid systems of control. The country had taken on a completely different appearance not long after the reform program began. China was bustling with economic development: Speed and efficiency became the main themes of a billion people's lives.

Lamentably, however, infrastructure projects such as power, transportation, and telecommunications lagged far behind. Especially telecommunications. In 1978, China had a switching capacity of only 4.05 million lines and only 2 million telephone subscribers; the telephone penetration rate was 0.38 percent, lower than the average rate in Africa, ranking below 120th in the world. A telephone was a luxury, exclusive to a minority in China's privileged classes. This was more than a century after Alexander Bell had invented the telephone.

The extreme inadequacy of communications facilities was a major bottleneck in the development of China's national economy. It became strategically imperative to build out large-scale telecom facilities, and rapidly. However, the country didn't have a single telecom equipment manufacturer that was worth its salt. For this reason, China adopted a policy of "exchanging the market for technology," which meant opening up its telecom market to foreign companies. Fortunately, the opening-up policy coincided with a global IT revolution, expediting the development of China's telecom infrastructure and allowing it to acquire the most advanced telecom technology at the time.

In retrospect, the decision to exchange its market for technology was timely and visionary. The automotive industry in China, in contrast, has suffered from market protectionism and has never really caught up with leaders in the industry.

It's painful to open the door, of course. When foreign companies entered the Chinese market, they charged high prices but offered little to no service. For example, a private branch exchange (PBX) line costs about US$10 these days, but in the late 1970s, they were US$500 a line. Chinese customers also had to wait lengthy periods for the product to be delivered and installed: The sellers were like kings. This is absurd logic in Western countries that boast long commercial traditions, but it prevailed in the Chinese telecom market in the 1980s.

At first, there were eight companies from seven countries (later nine companies from eight countries) that dominated the Chinese telecom market: Fujitsu and NEC from Japan, Ericsson from Sweden, Bell from Belgium, Alcatel from France, Siemens from Germany, AT&T (whose network division later became the independently operated Lucent Technologies) from the United States, Northern Telecom from Canada (later becoming Nortel after market consolidation), and Nokia from Finland. Most of them were already century-old companies at this time, each strutting into China like victorious conquerors, selling their products at prices that were sky-high.

The Destiny of a Generation

China rapidly developed its telecom infrastructure at considerable but necessary cost, and policy makers adopted strategies to develop the communications manufacturing industry at just the right time. Over 400 Chinese telecom manufacturers sprung up in the mid-1980s, including state-owned enterprises (SOEs), private companies, and companies with all different forms of ownership.

The dominant SOEs were Great Dragon, Datang, and ZTE. Born in 1953, the former military official Wu Jiangxing founded Great Dragon, later emerging as the "national hero of the telecom sector" and the "father of the Chinese large-capacity program control switch" before settling in as the major general of a military institute.

Datang was founded by Zhou Huan. Born in 1944, this former official of the Chinese Ministry of Posts and Telecommunications is now dean of the China Academy of Telecommunication Research under the Ministry of Industry and Information Technology.

ZTE was founded in 1985 in Shenzhen by Hou Weigui, who was born in 1942 and had previously served as a technology officer of the 691 Factory under the former Ministry of Aviation Industry of China. There is no doubt that each of these SOEs was committed to a national mission: to build up China's communications industry and issue a challenge to the "Western giants."

That's not to say that private companies didn't have a similar mission of their own. In 1987, Ren Zhengfei founded Huawei with the aim of becoming the backbone of the people's communications industry.

This was the shared destiny of his particular generation. Entrepreneurs born in the 1940s and 1950s embody many traits: a strong sense of patriotism, a bold splash of idealism, a sense of mission, integrity, a desire to lead, and an almost puritanical spirit of dedication. They were an outstanding bunch, with fire churning in their bellies and a penchant for rebelling against established rules. So whenever they were given the slightest bit of hope, they would choose to go out and challenge the world, even if the odds of winning were slim. The founders of the four major Chinese telecom manufacturers—Great Dragon, Datang, ZTE, and Huawei—are all cut from this very same piece of cloth.

In those days of reform that practically pulsed with unbridled enthusiasm, this generation of entrepreneurs radiated with an inherent vitality uniquely their own—the likes of which we rarely come across in China anymore.

Law of the Jungle: Survival of the Fittest

Looking back, one can't help but have the utmost respect for this old-school band of intrepid Don Quixotes. At their inception, China's four major telecom manufacturers were hungry for funding, technology, and talent. Even government-sponsored enterprises nearly came apart at the seams, as did Huawei, a private company with initial working capital of around CNY21,000. They were competing with the toughest rivals in the world.

The nine international communications equipment manufacturers doing business in China had operated for a combined total of 1,139 years. Eight of the nine were over a century old at the time. In comparison, these domestic telecom companies were like hundreds of newborn calves, new to the world and stumbling about, each going about their own business. They engaged in bloody competition with their Western counterparts, who by comparison were like strong, mature elephants. The word "bloody" here is no exaggeration: The law of the jungle is brutal, and the four major Chinese telecom manufacturers grew to adulthood atop the bones of hundreds of companies that the elephants had trampled to death. Even Great Dragon, the forerunner of Chinese telecom companies, eventually met its end.

But time brings seas of change. In only a few decades, the Chinese telecom industry has progressed by leaps and bounds, having ascended the ranks to join the United States and Europe as one of the top three in the world. During this period, China's domestic telecom companies grew strong through failure and defeat, defeat and failure, rising victorious in the China market—and then later in the global arena, startling their giant counterparts in the West. At the same time, mainstays of the international telecom market became fewer in number through merger and acquisition, or outright collapse. Only a few members of the old regime like Ericsson can still compete with unpredictable newcomers like Huawei and ZTE. In 2010, Huawei was listed among the Fortune 500, ranking number 397, the only privately owned Chinese company to make the list. In 2014, its ranking rose to number 285. And in 2013, Huawei surpassed Ericsson in sales, wresting away its all-time position as top dog in the industry.

Ren Zhengfei's prediction that Huawei would one day become one of the world's top three telecom manufacturers has come true. He is now focusing on how to create a balanced international business ecosystem and how to ensure a smooth rise as Huawei furthers its global expansion.

Huawei's Rise to Prominence

We have much respect for competitors like Huawei.
—Ericsson CEO Hans Vestberg

The story of Ren Zhengfei would be an amazing success story in the US.
—3Com Corporation president Bruce Claflin

The rise of domestic firms such as Huawei is a disaster for the multinationals.
—*The Economist*

This company is on the same road as Ericsson towards becoming a global giant. Now all these telecom giants eye Huawei as the most dangerous competitor.
—*Time*

A few years ago, a *Wall Street Journal* reporter asked John Chambers, CEO of Cisco Systems, "Which company worries you the most among all your competitors?" Chambers replied without hesitation, "Very simple—25 years ago, I already knew that my biggest rival would come from China. Now I can see it is Huawei."

Some of these remarks may seem a bit effusive, but no other company in China has earned such respect from markets in Europe and America, nor have they instilled such fear in their international competitors.

So how did Huawei do it? It took a combination of being in the right place at the right time. Twenty-seven years ago, Huawei was born in a residential building in Shenzhen as a limited company. This fledgling company had a meager CNY21,000 as initial working capital. According to the regulations of the Shenzhen government at the time (1987), a private technology company had to have at least CNY20,000 as registered capital and at least five shareholders. However, Ren Zhengfei had only CNY3,000 to his name. In order to obtain a business license, he looked for five other partners and pooled enough capital to satisfy registration requirements. In the years to come, the original partners were generously rewarded for their initial support, and Ren Zhengfei forged ahead on his own, setting the stage for the employee shareholding scheme. Twenty-seven years later, by the end of 2014, Huawei had 170,000 employees in more than 170 countries and regions, serving more than one-third of the world's population, and leading the world in international patent applications.

Huawei's success, in every sense, further testifies to the success of Deng Xiaoping's reform and opening-up policy.

In the mid-1980s, as times changed and the environment changed with them, policy makers in China weren't sure what form and direction the country's changes should take. Even Deng Xiaoping, the master architect behind the whole movement, couldn't produce a clear blueprint. So he proposed the idea of "feeling around for stones while crossing the river," that is, to move forward and figure things out along the way. He also pushed forward the theory that "it doesn't matter whether it's a white cat or a black cat; a cat that catches mice is a good cat," asserting that economic ideology is unimportant, as long as people contribute to China's development. In that particular political climate, although these guidelines were simple and expressive, they weren't easy to implement. In those days, without a bit of rash, adventuresome spirit—and even a touch of naiveté—it would have been difficult for that generation of swashbucklers to steel themselves for the plunge into private business.

Luckily, Deng Xiaoping was a firm believer in trial and error, encouraging everyone to go out and experiment with aplomb naiveté and if they made mistakes, just to scrap them and try again. When all cross sections of society were able to fully unleash their potential, the country's progressive reforms were hugely successful. China went from a planned economy to a planned commodity economy, then to a commodity economy, and finally to a market economy. Throughout this process, Chinese entrepreneurs rose up and fell back amid the dizzying twists and turns of evolving systems and regulations. Nevertheless, the goal of China over the past 30-plus years has been clear: to implement a market economy.

The ups and downs of its four major telecom equipment manufacturers perfectly exemplify the undulating period of growth during China's reform and opening up.

In the late 1980s, a group of senior-ranking R&D officials from the PLA founded Great Dragon, which became a shining star in the telecom sector. At the same time, Datang and ZTE were established under two central government ministries, and like Great Dragon, they were the "pets" of the country in the age of a planned economy. The government lavished funding and supporting policies on these SOEs and supplied them with vast amounts of talent. At that time, the stability and security of a job for life, dining from the "iron rice bowl" of a state-owned enterprise (so named because an iron bowl will never break), was the first choice amid job seekers, a trend that still persists among many today; only enterprises with close ties to the military were considered even better choices.

Private companies such as Huawei were marginalized and cast aside, their chance of success considered slim.

In the throes of incremental change, the system reached a tipping point, and major change arrived at once. After 20 years, Great Dragon collapsed. It simply disappeared. And Datang, equipped as it was with the advantages formerly granted to military and state-owned enterprises, lagged far behind due to the rigidity of its decision-making processes, HR policy, and incentive mechanisms, which prevented even the most daring and decisive company leaders from turning things around. Moreover, the telecom industry was becoming more liberal and global, and certain defeat awaited companies that couldn't adapt to change, innovate, and improve. Even century-old giants in the West began to suffer from the hardened arteries of inflexible management, and eventually began to fade.

ZTE was an exception. Although it was a state-owned enterprise, ZTE was based in Shenzhen, an experimental zone under China's reform and opening-up policy. As a result, ZTE reaped the benefits of both the old and new market systems: It received just as much government support and resources as Great Dragon and Datang, while enjoying far more freedom to innovate than state-owned enterprises based elsewhere.

And years later, its transformation into a state-owned, private-run company provided ZTE with a dynamic platform and major development potential.

Huawei is a different story. As a marginalized market castaway, Huawei would not have survived in the 1980s had it not been based in Shenzhen.

Adopting the principle of trial and error was perhaps the wisest decision made by the Chinese reformers from 30-odd years ago, because it drove institutional innovation and led to the establishment of a number of pioneering and adventurous pilot zones like the city of Shenzhen. This city abandoned the planned economy model, and dependence on interpersonal connections was far less necessary. A number of SOEs under central-government ministries began to build branches in Shenzhen in the hopes of reforming their own institutions. ZTE is an example of this, having belonged to the former Ministry of Aviation Industry. Naturally, private companies like Huawei also enjoyed ample freedom in this city.

For example, Huawei's ambitious equity-sharing program could not have been implemented and applied to more than 80,000 employees if the company had not been based in Shenzhen. Even in the early 21st century, such a shareholding plan was still defined as illegal fund-raising in some of China's inland provinces.

The fate of a country determines the fate of each individual and organization within the country. No one can steer completely clear of their country's trials and tribulations, nor can they grow unrestrained by the entanglements thus created. Before China's reform, Lenovo founder Liu Chuanzhi was too flexible and rebellious to confine himself to China's rigid market system, and the result was a series of frustrations. After the country's reform program was initiated, however, Liu harkened to the call of adventure and took the plunge, probing about in unknown territory and eventually building Lenovo, a business miracle of mythological proportions.

Perhaps to an even greater extent, Huawei is a miniature study of the country's progress. Founded by a former military engineer without any political background or social connections, Huawei has evolved into a world-class company through the dedication of hundreds of thousands of knowledge workers over 27 years. There are many secret ingredients in Huawei's growth story, but only two fundamental, indispensible elements: the changing environment and the creativity of its people.

Without a doubt, Chinas institutional reform was a major factor in Huawei's rise to prominence.

The Mysterious Power: Ren Zhengfei's Business Management Philosophy

The Founder of His Own Sect of Management Philosophy

Many have wondered how Huawei has succeeded, eclipsing most of its Chinese and Western peers, to become the world's largest company in its sector.

China's progress and its institutional reform were critical to the company's success, because Huawei couldn't have pulled itself up by its bootstraps if the environment wasn't right. At the same time and under the same conditions, however, more than 400 telecom companies in China have come and gone. Even in Shenzhen—a test bed for reform—five out of the six most valuable Chinese brands, except Huawei, started as SOEs.

From the 1980s onward, China was swept up in the largest wave of commercial development in human history. Businesses at the time were like ships, each raised up and carried along by the sheer momentum of the wave. Some, however, soon capsized and were swallowed up, while most drifted along, going with the flow. Others crashed against barriers in the sea or got stranded on deserted islands. Only a few rose atop the crest of the wave and survived, eventually sailing towards new lands. And thus the complex, indirect relationships between destiny and purpose, serendipity and necessity, opportunism and dedication, revealed their many faces.

Over the decades of reform and opening up, there have been countless stories of the rise, decline, and heart-wrenching downfall of private companies, their many carcasses lining the road of transformation.

Then how did a small dinghy like Huawei end up growing into a massive ocean liner? Is there any mysterious power or special arrangement of fate behind its ability to do so? Can its success be copied? A handful of politicians in the West remain suspicious of Huawei, and the media have tried their best to pry open the box. Both at home and abroad, many economists and management scholars are also extremely interested in dissecting this "specimen from the East."

Huawei now has over 170,000 employees, most of whom are knowledge workers. Questions about how the company has united and aligned such a large group of individuals abound. How has Huawei encouraged them to march forward as one, giving full play to their capabilities and potential? How has Ren Zhengfei—who can be likened to the company's spiritual leader, and who has only a 1.42 percent stake in Huawei—managed to exert authority within the company? And how has he retained it for more than two decades? The answer to these questions will unlock the secret to Huawei's success.

Western media have depicted Ren Zhengfei as the "godfather" of a business empire, as someone who has turned Huawei into some sort of religious following. This would be a clever comparison if it weren't meant in a derogatory way.

Ren Zhengfei has never been one to imitate or follow the example of others—it's simply not in him. For nearly his whole life, only solitude has kept him company. Although he has never explicitly said so, Ren has never fully accepted the conventional management theories touted in textbooks. Rather, he is the founder of his own sect of management philosophy. He is part meditator, deliberating over a single idea for months or even years, and part communicator, engaged in constant discourse with his management teams, outside experts, and customers, enriching his thoughts in an environment of open information exchange, letting his ideas ripen, and then harvesting them as fully fleshed-out systems of thought.

The Secret Ingredient to Huawei's Success

Huawei is a company of ideas. On countless occasions over the past two decades, Ren Zhengfei has shared his thoughts on the company's development—formally and informally, openly and privately, systematically and in fragments. His articles, speeches, and meeting minutes add up to millions of words. And the ideas that he has formed over the years are the very weapons with which Huawei has fought through the wilderness to rise in the international arena. This is what sets Huawei apart from most other Chinese companies, and even multinationals. More important than having ideas, however, is the fact that, in this intellectually curious organization, ideas can always be put into action.

To say that Huawei's success is rooted in its management philosophy is no exaggeration. Ren Zhengfei once said, "Without any special background or resources, we can rely on no one but ourselves; each bit of progress is in our own hands, and no one else's." He went on to note that the company's systems and culture are powerful forces—perhaps the mysterious power behind Huawei's success:

> I can tell you exactly what it took to unleash the potential of hundreds of thousands of Huawei people: it took 20 years of immersion in our management philosophy. Just as uranium atoms release enormous nuclear energy under neutron bombardment, every Huawei employee is their own atomic nucleus that, urged forward by our core values, has the potential to generate a staggering amount of force.

At Huawei, "business management philosophy" is top-level, metaphysical design, the prerogative of Ren Zhengfei and a select few others. They are the ones who must free their minds and think at an abstract level, waxing philosophical, leaving the world of concrete matters behind. They must be adept at contemplation, speculation, and gazing up at the stars, their minds projecting into the future.

"Methodology," on the other hand, is the architecture and construction of management systems. This is the domain of high-level managers. In this arrangement, the leaders are the visionary thinkers and managers are the builders, those responsible for materializing and even institutionalizing the philosophy of the company's leaders.

Thus far, Huawei has established a pyramid that is both stable and dynamic. On the very top of this pyramid are the thought leaders of the organization; next up is a group of business strategists and technology visionaries; the third level comprises hundreds of functional directors who have a strong grasp on strategy and can rally their troops, serving the role of commanders in their specific domains; further down are thousands of technology leads and middle managers in charge of operations, management, and R&D; and at the base of the pyramid are over a 100,000 knowledge workers, the true force and source of competitiveness at Huawei, driving their commercial chariot forward.

Looking back, poring over his old speeches and ideas, it seems that Ren Zhengfei's basic philosophy hasn't changed for over two decades. He summed it up himself:

> What is the driving force behind our rapid growth? It's a philosophy that's deeply rooted in the minds of our employees: it is customer centricity, dedication, and perseverance. Our success has nothing to do with any special background, and it's certainly not some supernatural power.

These have always been Huawei's core values. As time has passed, the company has enriched its core values with a management philosophy characterized by openness, compromise, and grayness. For over 20 years, Ren Zhengfei and his leadership team have mulled over these core values and management philosophy time and again, enriching them to form a complete system and cultural birthmark of sorts that distinguishes Huawei from other companies. They are the secret ingredients of Huawei's growth and success.

Understanding, Preaching, Practicing, and Sacrificing

The process of forming Ren Zhengfei's business management philosophy can be summed up in two words: understanding and mastery. Understanding a given philosophy is difficult enough, whereas mastering it is a whole other matter entirely. Mastery requires constant reflection on past success and failure, active reading, communication, and thought, and above all, putting oneself through the rigors of intellectual purgatory. Of course, self-control is a must. The mentally active and mentally ill are similar in that their brains are constantly firing on all cylinders. The difference is that the former can manage their delusions while the latter can't hold themselves back.

Ren Zhengfei managed to hold himself back. He has kept on thinking, but has never succumbed to psychosis. As his philosophical ideas are implemented, Ren advocates self-criticism as a way to optimize them. He believes that everything changes except change itself, so ideas have to change accordingly. Huawei's success can be attributed in large part to its culture of self-criticism, and the whole organization is committed to this aspect of its culture. From top to bottom, there are no exceptions—not even Ren himself.

In many religious traditions, the spiritual leader is responsible for spreading the good word. First and foremost, however, they're expected to be masters of their own doctrine before showing others the way. In a similar fashion, Ren Zhengfei is like the spiritual leader or "preacher" of Huawei's unique management philosophy—and by far the most captivating. His essays such as "My Father and My Mother" and "Spring in the North Country" have been published in dozens of languages. Any speech he delivers to internal staff exerts a subtle influence on their mindsets and priorities. And every time he connects with customers, politicians, and business leaders, he always leaves them with a profound impression of the company.

Ren Zhengfei is against running the business with meaningless slogans, preferring instead to lead his team with a system of values. Inside Huawei, core values are reinforced extensively at new employee and management training, during e-learning sessions, retreats, and business meetings. As a result, every member of a more than 100,000-person team, despite differences in background and personality, has the same set of core values pulsing through their veins. Such a broad and visceral diffusion of ideas is practically unheard of.

When it comes to spreading the word, Huawei is not a church, a philosophical salon, or a business school that stops at teaching ideas. Its business philosophy must be grounded, must take root, and must bear fruit. The application of philosophy to actual business operations requires painstaking effort.

At Huawei, philosophical doctrine and practical action complement each other. For example, the company advocates perseverance and dedication as core values, and it has adopted measures to ensure their manifestation. These measures include compensation packages that are well above industry standards, as well as other monetary incentives such as the Huawei's share-granting scheme in which employees directly benefit from the growth of the company. They include more opportunity-focused incentives too, like reserving promotions for dedicated employees (literally: those who keep fighting, who don't back down from struggle), and those who make the greatest contributions.

Gradual progress is also a trademark of Huawei's management philosophy. Ren has stressed on multiple occasions that conservatism is a good thing, and that organizations shouldn't be on a constant prowl for disruptive transformation and innovation. To ensure stable growth, an organization should stick to what has proven effective. While it may be less efficient, the costs of stability are also lower. Ren believes companies that innovate without purpose are chasing after the wind, and are likely to collapse at any time.

Huawei's management philosophy also stresses focus, what the company calls "strength in the face of solitude." These are fickle times, and staying the course without distraction is a massively difficult undertaking—especially in an environment that's suddenly full of new and tempting opportunities. Huawei has codified a number of restrictions in its company charter to check impulsiveness among decision makers, and thereby promote unwavering focus on the company's core business.

Operational mechanisms, including incentives, decision-making proce-dures, rules, regulations, and corporate culture, all play a critical role in the process of transforming business philosophy into actionable items. But a leader's commitment, courage, self-awareness, and dedication are also critical. Spiritual leaders should not confine themselves to the pulpit, after all, contemplating and preaching doctrine; they must step down, place both feet on the ground, and go out into the world, putting their beliefs into practice, sometimes sacrificing everything in the fulfillment of their holy creed.

Ren Zhengfei did just that, dedicating his entire life to Huawei—an awe-inspiring commercial edifice that took him 20 years to build. This sacrifice led Huawei to success, and has led Ren into a difficult life of solitude.

Solitary Founder, Solitary Company

Tearless Grief

Ten years ago, Huawei's chairwoman Sun Yafang was asked by a government official to describe how she felt about Huawei's success in one sentence. Without skipping a beat, she replied, "I feel like crying, but the tears won't come."

The official was shocked, completely at a loss for words.

Huawei was brought to this earth as a private company, and was therefore branded early on as a second-class company in China's old economic system. And although the system was constantly evolving, Huawei struggled to keep up with its twists, turns, and hidden traps. Ren Zhengfei has frequently noted, "Failure will one day arrive at our doorstep, so we must be prepared to welcome it. I've never doubted this—there are countless historic precedents." He also challenges the notion of success: "What is success, anyway? If we can survive after a lifetime of narrow escapes like Japanese companies did back in the day, then that is true success. Huawei has not succeeded; we have just been growing."

Comments like this are not a scare tactic. As a businessman, Ren Zhengfei must take risks—risk taking is in a businessman's nature. But on the other hand, he must always remain cautious, perhaps overly so, to avoid overstepping the wrong lines, which might topple the company. Day after day, year after year, he oscillates between a spirit of adventure and fear, placing him under tremendous pressure.

Before he reached the age of 44, Ren Zhengfei had lived a fairly normal life. Though his larger environment was inhibited, his personality was unrestrained. Throughout his life, his strong sense of idealism and self-motivation certainly made him stand out: He was elected as a delegate at the National Science Conference in 1978, and later at the 12th National Congress of the Communist Party of China in 1982.

After China began its reform, economic development became the zeitgeist of the period and Ren Zhengfei became a businessman. History shows that he chose the right path. Unfortunately—and fortunately, it can also be said—Huawei started from square one as a private enterprise. It was unfortunate because Ren had to navigate the company through institutional barriers, public criticism, and unfounded rumors, while dealing with commercial pressure and sieges from both domestic and international competitors alike. Huawei's growth was bogged down by a staggered chain of struggle; the company was constantly fighting to free itself from countless constraints and embark upon a path of upward development. The fortunate part was that this harsh environment forced Ren and Huawei to cast all illusions aside and explore their own path forward, a process through which they developed and codified a unique management philosophy and have attained tremendous growth.

Proof of Identity

Of course, Huawei's tremendous growth has come at a tremendous price. Ren Zhengfei once recommended a TV series called *Proof of Identity* to his management team. For years now, Huawei itself has been struggling to prove its identity—to give the world insight into what the company is all about.

Although it's a private company, Huawei has made great contributions to China's development—contributions that are no less significant than many state-owned enterprises. Over the past two decades, it has paid over CNY300 billion in taxes and has directly and indirectly created several million jobs. Together, Chinese telecom companies like Huawei, ZTE, Datang, and Great Dragon have saved the country trillions on infrastructure development. Just as an executive at a well-known Western telecom company once said, with its cost-effectiveness and disruptive technological innovation, Huawei has helped provide billions of people in the world with low-cost, high-quality information technology and services, which is valuable beyond measure. In China especially, where the information sector has evolved into a core strategic industry, the value of Huawei's contribution can't be denied. Despite its contribution, given the many resources that have been bestowed on state-owned enterprises and all the incentives that Huawei's foreign-invested counterparts have also enjoyed, Ren Zhengfei continues to wonder why the government hasn't provided *his* company with a bit more support.

Meanwhile, in the West, politicians and the media have yet to come up for air in their ongoing offensive against Huawei, making constant allegations of close ties to the military and the government. Some competitors have even sought to publicly demonize Huawei, claiming that it receives hundreds of millions of dollars from the Chinese government each year. In their hubris, many simply can't believe that a Chinese company without the "right affiliations" could show up on their doorstep in just a few decades' time, much less take a leading role in the market.

Huawei aims to establish its presence in every corner of the world. Naturally, Ren Zhengfei and Huawei have to prove themselves, over and over again. Perhaps no other company in the world has faced quite the same dilemma: In the international market, Huawei is seen as a representative of socialism, whose rapid growth might be viewed as a threat in the West; whereas in the Chinese market, as a private company, Huawei might be viewed as symptomatic of budding capitalism. Between this rock and the hard place, how in the world was Huawei able to grow?

Learning to Revel in Solitude

It comes as no surprise that Ren Zhengfei and Huawei are both solitary beings.

Contemplation, after all, is a solitary pursuit. And for over 20 years, Ren's contemplative interests have never strayed far from Huawei. From the very beginning, this man who started a company with almost nothing at the relatively late age of 44—an intellectual, former military engineer, and an idealist who could never bring himself to follow in the footsteps of another—set a practically impossible goal for himself and his company. It became a personal mission that demanded all of his time and energy. He has no hobbies other than reading and thinking. He has no friends in political, military, or business circles. In fact, he's been practically friendless throughout his school years, military service, and well into his tenure at Huawei. And yet he is a fantastic communicator, well versed and fluent in economic, political, and diplomatic affairs, both domestic and international. Above all he is an avid thinker, a contemplator, and a faithful preacher of his own ideas, which are all about the fate of his company.

Ren once said that it is relatively easy to maintain balance in a turbulent external environment, but internal environments are far more difficult to handle. Managing the divergent ideas and interests of over 100,000 people, with more and more joining every year, is easier said than done. This has troubled Ren and his colleagues for over two decades and, in their attempts to crack this problem, they conceived the company's management philosophy, which has been a deciding factor in Huawei's path to success.

But practical application is far more painful than pondering alone. Ren Zhengfei and his senior leadership team have had to rack their brains in order to push, motivate, and inspire more than 100,000 knowledge workers to take on new markets. And after rallying the troops, they have turned around and led the advance themselves, working closely with local teams in field offices around the world. From the ages of 44 to 70, Ren Zhengfei has kept his mobile phone on 24/7 and has spent more than one-third of his time traversing the world on business. Most of Huawei's senior executives, either still in office or retired, suffer from stress-related conditions like depression, anxiety, or diabetes and hypertension. Ren himself is no stranger to serious depression.

Death is an inescapable topic. In the end, the destiny of a business or organization boils down to what it can do to live longer. Huawei has survived for over two decades, while many of its peers have fallen, including industry giants that were once considered to be invincible. In these rapidly changing times, companies suddenly find themselves in the gloom of sharp decline, where only a moment before they were basking in the warmth of midday glory.

So who's up next? As the world looks to Huawei at the height of its success, will it be the next to fall? Is that how the Huawei story comes to an end?

Forty-four is a mysterious age. Four times four is 16, a number that symbolizes "a smooth road ahead" in Chinese. But ever since Ren Zhengfei founded Huawei at the age of 44, the road was never once smooth, and it doesn't look as if it's going to get smoother anytime soon. Will Huawei make it? Can it retain its depth of thought and the willpower to face any challenge head on?

These are some of the questions this book hopes to answer.

Common Sense and Truth: Customer Centricity

Chapter 1

Customers: The Only Reason Huawei Exists

Alcatel: The Lost Captain of Industry

One June day in the early 21st century, Alcatel chairman Serge Tchuruk received a Chinese guest at his private chateau in Bordeaux, France. The visitor was Ren Zhengfei, Huawei's CEO.

The sun was warm, and the endless estate of grapevines was packed with purple fruit that shone like gemstones. The air was filled with the unique charm of Bordeaux: quiet, proud, romantic, and dignified.

After they sampled two different wines, Serge Tchuruk got serious:

> In my life, I've invested in two companies: Alstom and Alcatel. Alstom is a nuclear power company. The business is very stable. The only variables are coal, electricity, and uranium. The technology does not change quickly and the market is not very competitive. In contrast, Alcatel is in a cutthroat market. You cannot predict what will happen the next day or the next month in the telecom market.

Ren Zhengfei could not agree more. The quiet and intoxicating air turned heavy and somber.

Serge Tchuruk is a widely respected industrialist and investor, and Alcatel was a leading telecom manufacturer at the time. After the dotcom bubble burst in 2001 in the United States, Alcatel was considered invincible, as were certain other European telecom companies such as Ericsson, Nokia, and Siemens. The open and liberal spirit in Europe had also nurtured a number of global telecom operators such as British Telecom, France Télécom, Deutsche Telekom, Telefonica, and Vodafone. While based and operating in Europe, they provided network services across all continents. American, Japanese, and Chinese telecom companies were far behind.

In the early 21st century, Huawei was still fighting an uphill battle. Ren Zhengfei was shocked when he heard about the troubles of Alcatel, the captain of the industry. After he got back to China, Ren shared Serge Tchuruk's views with Huawei's senior executives and asked them, "Does Huawei have a future? What is the way out?"

Huawei then became immersed in a great debate. A consensus was reached that the company must continue to embrace "customer centricity." The underlying principle that had taken the company so far would remain fundamental to its future—that customers were the only reason for the existence of Huawei, and, as a matter of fact, of any company.

Of the four strategic statements that Huawei later developed, the first one reads: "Serving customers is the only reason Huawei exists. Customer needs are the fundamental driving force behind Huawei's growth."

Turn Your Eyes to Your Customers and Your Back to Your Boss

On a flight from Shenzhen to Beijing, a passenger about 60 years old sat in the last row of the first-class cabin. Throughout the three-hour journey he was reading quietly. This passenger was Ren Zhengfei. When the plane landed in Beijing, he stood up, picked up his luggage, and joined the crowd just like other travelers. He didn't have an entourage and no one picked him up. When he travels for meetings or other purposes, he usually doesn't inform people from local offices. He simply takes a taxi straight to the hotel or to the meeting. He is used to taking taxis, which occasionally makes headlines when he is spotted.

This is also the custom of most other senior executives at Huawei.

A deputy chairman of Huawei said, "This habit does not mean that Huawei's leaders are all more sensible than others; that's not why we do it. It's important because it's a manifestation of one of Huawei's core values: Are customers important, or our executives? This is the founding principle of right and wrong at Huawei, and is critical to the company's success."

Ren Zhengfei also cautioned many times:

> There seems to be this air in the company that people's bosses are more important than our customers. Perhaps because management has too much power, some people care more about their bosses than their customers. They develop fancy PowerPoint slides to present to their bosses, and they make extremely thoughtful arrangements for executives traveling on business. The question is: If they spend so much thought on this, how much is left over for our customers?

Ren, therefore, commanded:

> Everyone in the company must turn your eyes to your customers and your back to your bosses. Do not go crazy doing slides to impress your bosses. Do not assume that you will get promoted just because your boss likes you. If this happens, it will weaken our ability to fight.

At a meeting in 2010, Ren Zhengfei proposed:

> Huawei should and will promote those who turn their eyes to their customers and their back to their bosses. For those who turn their backs to their customers while focusing on their leaders, we will absolutely let them go. The former are value creators for the company, while the latter are just a bunch of lackeys who are in it for personal gain. Managers at every level of the organization should recognize the value of employees who turn their backs on you. You may feel uncomfortable at first, but they are the right people for the company.

Huawei is a typical private company in China. It started as a trading company and is therefore intimately familiar with the path to success as a vendor. It is committed to serving its customers, and holds true to the maxim that "the customer is king." It's in the company's blood and has helped it tremendously when it began to develop its own products. Although its products were not as good as those of its competitors at first, Huawei was still able to win its customers' understanding and support through its excellent service. Later, when its products became as good or even better, customer loyalty came naturally. A firmly held belief that customers always come first has essentially served as Huawei's guiding light over the past 27 years.

Who am I? Where do I come from? Where am I going? These are the fundamental questions all companies have to answer. Without fail, every successful company can answer them correctly, although the answer may vary in different periods of development. Some enterprises embrace customer centricity from day one, but after an initial bout of success, they end up losing sight of who is most important: their shareholders, their employees, their customers, or their managers? Different answers to this question reflect different attitudes in the pursuit of success, and eventually, they determine the fate of the company.

Magic does not exist and truth can't be reversed. The alchemy behind Huawei's growth simply lies in its core values: customer centricity, dedication, and perseverance. They are plain and unadorned—nothing deep about them. However, from the day Huawei was founded, Ren Zhengfei and his colleagues have believed this to be true, and these three values have imbedded themselves in the very bones of Huawei's culture. They have become the tenet for each and every member of the organization.

In December 2010, Ren Zhengfei gave a lecture titled "Embrace Customer Centricity, Value Dedicated Employees, and Remain Long Committed to Dedication" to the senior executives of a large European telecom company. He told the audience:

> This is the entire secret behind Huawei's ability to surpass its competitors, and this is the guarantee for Huawei's continued success. It's not a message of foresight, but a lesson we have drawn from practical experience. The three pillars are interconnected and support each other. Customer centricity sets the

direction for dedication; dedication is the means to customer centricity; and dedicated employees drive long-term commitment to customer centricity.

From this speech it is clear that the "soul" of these three pillars is in fact "customer centricity." Without it, what would be the direction and meaning of dedication? Without customer centricity, how to evaluate the value created by the dedicated employees? Not to mention distributing that value back among them.

Distorted Common Sense

Customer centricity was not invented by Huawei; it's a universal business value. The idea that "the customer is king" was invented in the West and has permeated its business history. The idea is very simple: Businesses are about making money and companies that fail to generate profit are worthless. And where does the money come from? From customers, of course. Therefore, the companies with the most potential are those who can find a way to make more of their customers dig willingly into their pockets as deeply as possible, for as long as possible. This business philosophy in the West has evolved for centuries around a permanent theme: The mission of a company, its managers, and its products is to satisfy customer needs.

This is a plain fact. A business has to pay taxes to the government, pay its employees, and pay its suppliers for goods; customers are the only source of incoming money. Yet, no customer would give the business money for nothing. Customers are rarely that altruistic. They have the right to choose and will only pay companies who provide high-quality products or services, and at a reasonable price. In this sense, you can't go wrong with customer centricity.

Therefore, management expert Peter Drucker believes that the purpose of a business is to create customers.

With the rapid development of the securities market, capital has been the king of the world for 30 years, and traditional business ethics based on customer centricity have been wildly distorted. In the US market, for example, it's now a common business ethic to maximize shareholder value. Businessmen are focused on the stock market and make business decisions based on the opinions of securities analysts. As a result, companies might expand rapidly or burst overnight. They might become industry giants in three to five years, or see their market value plummet within days or even hours. The Chinese market is no exception. Industrialists become rising stars in the capital market, but their companies inflate and burst like balloons. They are crazy opportunists; as for their companies, they either burn bright and burnt out fast, or struggle for breath in an endless chase to keep up with their shareholders' short-term expectations.

Common sense has been distorted. Customer centricity, once common sense in the business world, has become the solitary pursuit of a select few leading enterprises.

Who Will Feed Us If We Forsake Our Customers?

Huawei is among the few that seem to have maintained a grasp on common sense. Huawei's management believes that the company lives and dies by its customers' value chains, and that its own value depends on the value of the whole chain. In other words, Huawei derives sustenance from its customers—it would starve if it didn't serve them. Huawei can only stay alive and prosper with customer centricity as its guiding force.

In July 2001, *Huawei People*, the company's internal publication, featured an article titled "Serving Customers Is the Only Reason Huawei Exists." The title was originally "Serving Customers Is the Reason Huawei Exists," but Ren Zhengfei revised it. He believes that Huawei was born to serve its customers and that the company has no other reason for existence. Serving customers is the only reason.

Qian Zhongshu, a renowned Chinese scholar and writer, once said that "truth is naked." Huawei's executives say something similar: "We must not complicate Huawei's culture, or dress it up with too many details." Customer centricity is the very essence of commercial activity. A company can only survive if its customers are satisfied. This is a simple truth, and yet it is extremely difficult to put into practice. Whoever can stick to it will be the winner.

Huawei was once accused by some Western companies of having destroyed the telecom industry's value chain. Huawei's Deputy Chairman of the Board and rotating CEO Hu Houkun responded this way:

> They only saw the price of certain products drop, but overlooked a basic law of value: that is, when prices drop to a critical point, there's an avalanche effect in customer demand, growing explosively—many thousand times fold. On the one hand, Huawei's contribution was to reduce the purchasing cost of its customers. After prices reached a reasonable level, we helped spur the potential demand for information services among billions of people around the world, thereby benefiting the entire industry.

In fact, it was Huawei's global expansion, combined with its customer centricity that, in a way, helped expedite the human race's entry into information society by at least 20 to 30 years across multiple dimensions.

Professor Huang Weiwei, a management advisor to Huawei for 20 years, once told a story: A group of EMBA (executive master of business administration) students from the School of Business at Renmin University of China visited Lancaster University in the United Kingdom. They were deeply impressed by the UK's glorious industrial history as well as its modern development. One of them mentioned Huawei to their British professor.

The professor said, "Huawei is walking along the same path that some of the world's most remarkable companies once travelled. Before they reached the top, they were all customer-oriented, completely dedicated. And yet when they reached their peak, they became complacent, stop listening to their customers, which marked the beginning of their eventual decline."

But it's the plain and simple truth. Huawei does not believe in short-term economic magic. While competitors such as Ericsson and Motorola create plans for each financial quarter or year, Huawei plans its development by the decade. According to Xu Zhijun, Deputy Chairman of the Board and one of the company's rotating CEOs, this is one of the secrets behind Huawei's ability to catch up with and surpass its competitors. If Cisco and Ericsson had not gone public and had focused entirely on their customers, Huawei wouldn't have been able to get close to them.

Of course, public companies don't necessarily ignore their future or their customers. Apple is a success story. When Steve Jobs died, the whole world lamented the loss of a man imbued with the true spirit of innovation. However, to be more precise, Jobs was a great listener and was superb at integrating resources. Apple's products are characterized by simplicity plus aesthetics, which are precisely what consumers want. A simple life is good, but a bit of art makes life even better.

Fundamentally speaking, technology and products are not at the heart of Apple's success, but rather the humanistic spirit of Steve Jobs and his manic sensitivity towards customers. The question is, without Jobs, who will keep the company focused on customers and who else has the natural capacity to see directly into their hearts? Under Tim Cook's leadership, Apple will face great challenges.

The Fall of "Century-old Empires"

An Age Where Change Has Outpaced Our Ability to Adapt

The history of business since the 1990s has been brutal, verging on downright gruesome. If the Industrial Revolution had started with violent plunder and nakedly cruel conquest, the commercial world today, in an age of globalization and informatization, is also engaged in a war that is no less bloody. This war is even larger in scale and more catastrophic. In short, the past two decades have been one giant debacle. The telecom industry, for example, has been basically overturned, and then overturned again. This veritable rollercoaster of disruption and being disrupted has shaken those involved—both participants and witnesses alike—down to their very core.

Take, for example, Lucent Technologies—named by *MIT Technology Review* as the world's best technology company in the telecom field for two consecutive years. In January 1999, the company had a total market value of US$134 billion and its share price was at a record high of US$84. But in September 2002, its share price was no more than US$1 and their stock was rated as junk. In its peak years, the company had 153,000 employees, but by this time they had been reduced to 35,000. Eventually, Lucent Technologies was merged with Alcatel in 2007. However, the Alcatel–Lucent conglomerate did not fare well. Post-merger Alcatel–Lucent suffered from declining income and continuous losses for five consecutive years, amounting to US$12.4 billion. They were forced to cut jobs. Worse yet, the company was implicated in financial scandals and cultural conflicts. They went from top to bottom faster than lightning itself.

Motorola shares a similar story. The company was once hailed as a money printer in the telecom industry. Established in 1928, Motorola distinguished itself as an innovative pioneer. It developed the world's first prototype cellular phone and the first commercially available mobile phone. Motorola is unquestionably the father of mobile phones. But the market was relentless. In its peak year of 2001, Motorola had approximately 150,000 employees, but by the end of 2003, this number had dropped to 88,000. At that same time, Huawei had only 30,000 employees. Today, Huawei has just as many employees as Motorola had at its peak.

An authoritative figure in the Asia Pacific telecom scene once commented that the 1990s were a "Neolithic Age" in the history of human telecom, and that whoever took hold of that decade would take over the world: "If it was not for those 10 years, Motorola would have bit the dust long ago. That decade not only saved a bunch of old Western companies such as Motorola, but also cultivated Huawei—a company severely overlooked by the West. For a long time, the Americans never viewed Chinese companies as their competitors. In their eyes there were only European and Japanese ones...."

On October 15, 2011, Google bought Motorola Mobility at a price of US$12.5 billion, and almost at the same time, in 2012, Huawei decided to recruit 28,000 new employees. In 2012 and 2014, Google resold its set-top box and mobile handset business—both outside of Motorola's core patents—to Arris and Lenovo respectively.

Does that mean Motorola has reached the end of its line? Or will it reenter the sector as part of an Eastern company? Only the future can tell.

There is no secure place in the world. In 2006, Finland's Nokia and Germany's Siemens announced that the companies would merge their telecom equipment businesses to create one of the world's largest network firms, Nokia Siemens Networks. Several years have passed and this mammoth has also suffered the unfortunate fate of its species, having served as proof that one plus one is in fact less than two. By the end of 2013, Huawei took over Ericsson's position as number one and became an equal leader in the global telecom industry.

In April 2015, Nokia announced a plan to acquire Alcatel–Lucent for €15.6 billion. Analysts predicted that the global telecom suppliers' scene would see a competition between three great powers: Huawei, Nokia, and Ericsson.

As a Chinese saying goes, every generation has its heroes, and each may lead the way for decades. These days, however, influence can die off after only a few years. Spring flowers bloom both strong and bright, and yet the ground is littered with yellow leaves, almost overnight.

An age in which change outpaces our ability to adapt has definitively arrived.

Will of Capital or Will of Customers?

Many analysts believe that the fall of Lucent Technologies was owing to the bursting of the telecom bubble in 2001. Some scholars attribute the decline of Motorola to its arrogance, which led to bad decisions and the inability to make a comeback.

They are right to some extent, but they haven't touched upon the key question: What were these companies' core values? Were they based on customers or something else?

Paul Galvin, the founder of Motorola, had never believed that profit is the sole or highest aim of the company. Instead, he believed that the mission of the company was much nobler than making a profit. Likewise, his son Bob Galvin, when facing the challenges brought about by Japanese semiconductor companies in the 1980s, also stressed that efficiency wasn't what was most important, but rather having a good grasp on the future. As a result, Motorola was able to turn out a revolutionary product that catered to customer needs: commercialized mobile phones. According to Bob Galvin, the key to the success of a company was everyone working towards the same goal and proactively serving its customers. This is exactly why Motorola was able to go from good to great. But when the company deviated from this principle, it inevitably slipped from greatness into mediocrity.

In 1991, Motorola decided to invest in a global communications system: the Iridium Project. Yet, the project survived for less than a decade. In March 2000, Iridium Satellite LLC went into bankruptcy, with Motorola incurring a loss of US$5 billion. Iridium Satellite was once highly acclaimed as a revolutionary communications tool, but was hindered by a lack of subscribers and a prohibitively high tariff. The project had not moved along and this was a watershed for Motorola. At the time, Motorola had not only grown insensitive to customer needs, but had also confined itself to a closed-off ivory tower of its own creation. The company rejected the commonsense position that the customer is king, instead believing in the omnipotence of technology.

Many other companies that failed or died out behaved in the same way. Established in 1996, Lucent Technologies was one of them, and its fall was even more shocking. It was formerly a division of AT&T and had been a favorite among investors from the very beginning. The influx of investment, however, had pushed the company into a path of utilitarian expansion. Within six years, the company acquired 36 businesses with the aim of rapid growth. Lucent was engaged in a profit-making rush to satisfy the endless desire of the capital market. The result was a floating organization with diverse and incompatible cultures, and finally came the shocking debacle.

Lucent Technologies was torn between utilitarianism and idealism. On the one hand, it had to cater to the stock market, trying its best to keep up with the pace of expansion. On the other hand, it had the best research lab in the United States and even in the world—Bell Labs. Producing 12 Nobel laureates, Bell Labs had been the pride of the United States and a global leader of scientific research and development since the early 20th century.

In 1997, when he visited Bell Labs, Ren Zhengfei was deeply impressed. Stirred with strong emotion, he said, "I was told more than 10 years ago that Bell Labs could produce one patented invention a day, and now it can produce three inventions every day. That's so amazing. My feeling for Bell Labs is stronger than love."

However, as a core asset of Lucent Technologies, Bell Labs suffered serious conflict between the venturist culture of the capital market and the tradition of scientific research. Bell Labs traditionally put great emphasis on basic research, but now, quite inevitably, had to focus more on pragmatic, market-oriented inventions. The organization has gradually lost its future-oriented idealism.

In fact, neither the profiteering of the capital market nor an idealistic commitment to the future is fitting for the true mission of a business: Neither approach is focused on the customer. Therefore, Lucent Technologies' fall from greatness seems to be all the more logical in the business world.

Opportunism: The Golden Cup of Poisoned Wine

Taking risks is a common characteristic of entrepreneurs. But taking risks is not the same as gambling, which is a game of intellect that relies on cleverness, intuition and luck. Ren Zhengfei once said that everyone has the inclination to gamble, but that he had to control himself. He said he had to control such inclinations and learn to manage them. When he was in Las Vegas, he would visit the casinos as well as anyone, but he never placed any bets. He would just hang around. He said with a smile, "I'm afraid of getting sucked in. . . ."

A man called Wu Ying is just the opposite. Ying in Chinese means hawk, and this name is not far off from his personal disposition. He has sharp eyes, and with them he spots the very best opportunities. He was the one who bought the Personal Handyphone System (PHS) technology from Japan at just the right moment when China decided to open up its communications market to competition. As a result, he won an almost impossible game, and a company called UTStarcom rose in the market. PHS was considered outdated technology at the time, so leading international telecom companies, including Huawei, did not believe that the technology had any prospects. However, the Chinese market had a strong demand for this system. PHS was adopted by China Telecom, which had not yet obtained the license to operate mobile phone services. PHS, as an extension of the fixed-line connection, fits in well with the ability to provide both fixed and mobile services.

Owing to its adventurism and speculation, UTStarcom quickly rose to stardom. In 2003, UTStarcom occupied over 70 percent of the device market. When the PHS market slowed down in 2005, it still maintained a 60 percent share of the systems market and 50 percent of the device market. Its 2003, its sales revenue was about on par with Huawei.

UTStarcom's PHS was a huge gamble, causing excitement in the capital market. In 2000, UTStarcom was listed on NASDAQ, and in the same year, the company made it into the Fortune 1000. For 17 consecutive months, its performance exceeded the expectations of Wall Street analysts. Sales ballooned, growing by a factor of 100. In October 2003, *Business 2.0* magazine ranked the company number one among the 100 fastest-growing technology companies of the year. Wu Ying, the company founder, became an overnight celebrity in the telecom sector.

This momentum, which on the surface was highly positive, ended up spoiling UTStarcom. Opportunism is like a coin with two sides: Luck has it that opportunities don't usually fall to the same company or person over and over again. A company has to depend on its strength and capability to succeed. On June 1, 2007, International Children's Day, Wu Ying resigned from UTStarcom. A company which had grown to a size that belied its age was about to sink.

Putting the market first is not the same as putting customers first. A company should not just follow trends or simply cater to short-lived market opportunities. Great companies like Apple and IBM have stuck to the creed of customer centricity. They are committed to satisfying both their immediate needs and their potential longer-term requirements, and therefore offer customers the best products and services. Luck in the market is a golden cup that is sometimes filled with poisoned wine. Anyone who drinks from the cup becomes shortsighted, lazy, speculative, complacent, and insensitive. Even their muscles begin to atrophy.

A company that pursues excellence and long-term development must not depend on mere luck.

The Hand That Holds a Torch against the Wind Is Certain to Get Burned

Society has entered an era in which stars tend to burn up quickly, while outbursts of short-lived meteors crowd the midnight sky. New technologies emerge in waves as the Internet threads the world together like a spiderweb. All this is destructive to tradition and has caused a global organizational crisis. Systems of organizational theory and practice developed over several thousands of years now stand at the cliff's edge, with business organizations affected most.

Over the past 30 years, many companies have rapidly emerged and disappeared. Names like Wang Laboratories and Yahoo entered the stage from nowhere, only to swiftly fall into decline. Some time-honored companies, likewise, have turned from stars into meteors in this period of dramatic change: AT&T is a perfect example.

AT&T was a truly prestigious player in the global communications sector. Established in 1877 by Alexander Bell, father of the telephone, it gave birth to many distinguished organizations, such as Bell Labs, Northern Telecom, and Lucent Technologies. It also produced Claude Elwood Shannon, the father of information theory, and a number of Nobel Prize, Turing Award, and Claude E. Shannon Award winners. After two antimonopoly movements in the 20th century, it hadn't lost its edge and had, in fact, become more competitive. Regrettably, it was unable to survive new technology and the capital market that hit the world like a storm in the early 21st century. The AT&T family withered away, and these days it isn't much more than a shiny shell of its former self.

This seems to be an unavoidable fate for telecom companies: If they are committed to innovation, they destroy themselves; if they stay away from innovation, others will kill them.

There is yet another fate: Willingly or not, they end up in the slaughter-house of the capital market. Wall Street, for example, is well known as the butcher of the world's real economy. Financial capital incubates a number of tech start-ups with astonishing speed, but kills them just as quickly. Not even AT&T's business empire could escape this fate. It was rent into three parts and abandoned after being stripped of all its flesh. Its ending is as gruesome as most other stories in the capital market: Capital and professional managers, or capital alone, are the only winners, and companies are left to suffer or die out entirely.

In this globalized world, worship of capital and technology has destroyed a number of great companies and entrepreneurs. On the dead or critically ill list are Wang Laboratories, Motorola, Lucent Technologies, Northern Telecom, AT&T, and Yahoo, to mention just a few. Motorola and many other public companies are nothing but marionettes strung on the golden fingers of financial capital these days.

Over the past two decades, Huawei has had many opportunities to wed itself with venture capitalists, but Ren Zhengfei chose to avoid them.

Huawei hasn't joined the series of short-lived meteors because it has resisted the siren call of the capital market. The capital market can quickly fatten up companies and entrepreneurs alike, but can just as quickly destroy them and their delusions of success. As the Buddhist precept goes, "Desire is like holding a torch against the wind: The hand that holds the torch is certain to get burned."

In other words, the leadership of Huawei has kept a clear mind: The underlying factor for the company's sustained success is its customers, rather than technology or capital. This is a core value of the company. What Huawei needs is a customer-oriented culture, not one that rubs shoulders with capital investment.

Customer Centricity: The Result of 27 Years of Contemplation

1987–94: Survival Was Everything

If you travel across the canyons of Mount Wuyi, you will be amazed by its tranquility, all the chirping birds, and the fresh air. You may also notice countless ants carrying dirt from the bottom of the cliff to small clefts higher up the mountain, where they use the dirt to build their nests. They are busy working to prepare for the cold winter or the rainy season. If they fall to the ground, they start over again, but if they happen to fall into the flowing stream, they are washed away.

This is, to some extent, symbolic of the fate of many small- and medium-sized enterprises (SMEs) in China.

SMEs play a significant role in China's economy. There are now over 40 million such enterprises in China, most of which are privately owned. They account for 60 percent of the GDP, 80 percent of all jobs, and 50 percent of China's tax revenue. However, they are rather short-lived. According to the statistics, the average lifespan of Chinese enterprises is less than 2.9 years, and each year, more than a million go out of business. As of 2010, the average lifespan of Chinese SMEs was 3.7 years; in Europe and Japan, it's 12.5 years, while in the United States it's 8.2 years. Since the beginning of China's reform and opening-up program, there have been a number of Chinese companies cited as successful cases for MBA students, but by now 80 percent have fallen. In other words, people are rushing toward the glass door regardless of risk, but most end up seriously hurt and some even lose their lives. They go through the process of registration to cancelation in a short period of time, with their start-up investment leading to nothing but a mess and a body covered in scars.

Survival is a tough task for companies, as Huawei chairwoman Sun Yafang once said. It is especially daunting in the company's infancy. According to Ren Zhengfei, Huawei's success is attributable more to market opportunities than the company's capabilities.

Huawei started as a trading company with a dozen employees. This was a poorly equipped team: They had neither product nor money. However, they fought through the lines of foreign companies and Chinese state-owned enterprises and found a hope of survival. At that time, Huawei's slogan was, "Let us drink to celebrate success, but if we fail, let's fight tooth and nail to come out alive." In this specific context, success meant survival. Ren Zhengfei was never a mere merchant aiming at making money. Yet reality was cruel: He had to make enough money to survive and get stronger.

At Huawei, suffering is nothing in the pursuit of ideals, but it's unbearable when your ideals turn into a joke. At the very start, Ren Zhengfei called on his colleagues to build a world-class enterprise. Most would rather believe that this was an honest pursuit, but he himself was clear from the start that this was a mission impossible. To accomplish world-class repute, the company should first struggle to survive and live long enough. Therefore, from the first day of the company, Ren Zhengfei and Huawei came to embrace a proper fear of death. Survival became Huawei's most basic—and perhaps most lofty—strategic goal, and pragmatism was its only choice.

A review of Huawei documents and speeches from Ren Zhengfei before 1994 would unearth terms like "wolf and Bei[1] spirit," "call for heroes," and "to be shameless" ("shameless" here means to possess the nerve to face customer complaints and the courage to challenge oneself). Customer centricity was hardly mentioned, let alone discussed at length.

During that period, Huawei was still one of millions of ants struggling for survival in the Chinese market. It was a small trading company with ideals, but without values. Its ideals were abstract, and it could not define its values because just surviving was a challenge.

Customers Are the Heart and Soul of Huawei

Starting in 1994, Huawei started to outgrow the chaos. After groping for stones in the river for seven years, fighting in the jungle for survival, Huawei began to put on some muscle. In October 1994, Huawei launched

---

[1] Creatures in Chinese mythology that are almost exactly like wolves, but with extremely short front legs and long back legs.

its first telephone switch system, the Huawei C&C08. It was a milestone in the history of the company because it marked the end of its trading days without products or technology to call its own. Huawei had entered a new era.

In June 1994, Ren Zhengfei delivered a victory toast in which he said:

> We are faced with price pressure as the market fills with both good and bad products. But I believe our sincere commitment to customer service will move our "kings." They will know that our products are worth the money they pay. This will ease our difficulties, and we will surely survive.

In a speech in 1997, Ren Zhengfei said:

> Huawei is a profit-seeking entity and everything we do is to create commercial value. Therefore, our culture is called corporate culture, not any other culture—and certainly not politics. Huawei's culture is a culture of services, as only through services can commercial value be created. Service is a broad term. After-sales services, product research, production, upgrade, and optimization are all part of it. It has to steer the way we build our teams. If one day we no longer need to serve, we'll have to close the company. In this sense, service is the lifeline of our company and for us as individuals.

For a long time, "Huawei" meant low prices, poor quality, but excellent service. A Chinese domestic operator still clearly remembers the following:

> In the early years, Huawei's switches were used mostly by county-level telecom operators. They were not very reliable and often broke down. But Huawei did a very good job at providing service; its people were available around the clock. At that time, the staff of telecom operators fell into the habit of bossing people around, often scolding Huawei's staff, and even Ren Zhengfei himself. None of the employees would argue with their customers; instead, they would sincerely apologize and get the system up and running again as soon as possible. This was in great contrast to Western companies, who were used to blaming the customers for problems, and were insensitive to customer

needs. As a result, Huawei left people with a deep impression. How can you refuse anyone who really treats the customers as king? In the 1990s, service was a rare concept in China, but Huawei had made service into an art form.

Nevertheless, this was still one-dimensional thinking in an age of trade, an outcome of forced acceptance and cognitive restraints. It wasn't until 1997 that Huawei explicitly introduced its concept of treating customers as the company's foundation and looking towards the future. Ren Zhengfei explained, "If we are not tuned into the needs of our customers, there would be no foundation for us to exist; if we are not oriented towards the future, there would be no traction for us to move ahead, and we then slack off and lag behind."

Since then, with some slight changes to his wording, the concept of customer centricity has become the guiding principle for every activity of the company at every stage of its development. In 2002, Ren Zhengfei said:

> Customers are the heart and soul of Huawei. As long as we have customers, Huawei's soul will always persist, no matter who leads the company. A company is very fragile and will run into trouble sooner or later if it depends too much on one leader.

In 2003, Ren added:

> We've always stressed that we must stick to customer-orientation. This should be a rational, unequivocal, and unforced orientation that represents universal truth in the market. Any forced, vague, or policy-mandated demand is not a real customer need. We must be able to distinguish real needs from opportunistic demand, and maintain a rational approach to meeting customer needs. And we must not rule out adopting different approaches at different times.

On an existing foundation of customer centricity, Huawei executives further deepened their understanding of the company's approach, the premise that Huawei should orient itself in a "rational, unequivocal, and unforced" way towards customer needs, a way that represents universal truth in the market. It was with this understanding that the company refrained from investing in PHS. This led to a heated debate within the company.

At the time, had Huawei invested CNY20 million and 30 key engineers, in six months they could have churned out products that would bring in around CNY10 billion. The company's top management, however, went against external and internal pressure, choosing instead to focus on the research and development of the WCDMA (Wideband Code Division Multiple Access) technology. Helplessly eyeing the rapid rise of UTStarcom and other domestic companies, and at the same time suffering the company's first bout of negative growth, some Huawei employees started spreading doubt within the company. But Ren Zhengfei's position remained firm: Huawei would not pass up any business opportunity, but it was also a company with lofty ideals.

Customer centricity is a general concept, but it has concrete meaning for Huawei: not only to meet customers' current needs, but also to build long-term customer satisfaction. In 2014, when Russia was facing a new economic crisis, someone suggested that Huawei lower its product quality to ensure lower costs. Rotating CEO Guo Ping immediately declined the suggestion. To him, this was against the essence of customer centricity: "We should not aim to simply please our customers. To be accountable for customer satisfaction, we need to take their perspectives into consideration, and then draw clear lines for product quality and innovation that we will not cross." This would be truly sustainable development.

Since 2005, Huawei's relationships with its hundreds of customers have grown beyond the purely transactional. They have become symbiotic and mutually enhancing strategic partnerships. This was a fundamental change for Huawei. But companies at this point are even more likely to lose their grasp on common sense. Their concept of customer centricity, as well as their system of values, may become distorted. It was precisely at this stage that some Western companies began their downward spiral in spite of rosy prospects. The decision makers at Huawei stood in awe as they bore witness to the collapse and fall of these companies like so many mountains caving in on themselves. And through this process they came to realize the incontestable strength of common sense and essential truth. As a result, between 2006 and 2010, Huawei exhorted customer centricity widely throughout the company and organized a series of training sessions to reinforce the idea at all levels.

The minutes of an Executive Management Team (EMT) meeting in 2010 clearly state:

> Our long-term strategy is to make the company more competitive in the market through improved quality of products and services, and enhanced delivery capability. This will help us develop and maintain a balanced position between ourselves and our peers from the West. Competing on price, not on service quality, would not place us further ahead of the strategic competitors, while we might pressurize the room for development of Western vendors.

This was a new interpretation of customer centricity after Huawei began building strategic partnerships with its customers. Huawei also changed its approach towards its competitors, and its market space has thus been extended.

In 2010, "customer centricity, dedication, and perseverance" were officially incorporated into Huawei's core values.

"The Misplaced Winner" and Huawei's Position on Innovation

"How could they possibly beat us?"—While some people in the West were expressing astonishment over Chinese companies like Huawei, David Wolf, an American consultant and industry analyst, answered this very question. In his book *Making the Connection: The Peaceful Rise of China's Telecommunications Giants*, he gave the following warnings to his peers in the West:

> The telecommunications sector is the first of China's industries to step out and challenge global giants. And it is important that we get into the habit of doing so, because by lumping all Chinese industries and companies together into a single mass and painting them all with a tarred brush would not only deprive us of opportunities to work with, learn from, and prosper with the best of these companies, it would also build a wall of intolerance between China and the West that the world can ill afford….

David pointed out that, after the establishment of the new China, there existed two latent but important elements: opportunities and entrepreneurs. He stresses this proposition in the introduction of his book: The historical opportunities brought about by China's reform and opening up, coupled with a bunch of adventurous, visionary entrepreneurs such as Ren Zhengfei and Liu Chuanzhi, created the right "time, place and people" that led to the development of Chinese multinationals such as Huawei and Lenovo.

After mountains of research, Wolf came to another conclusion: In the competition against telecom giants in the United States, Europe, and Japan, the "final winner" shocked almost everyone—the companies expected by the government to lead the industry had all failed, but "a few tenacious local players like Huawei survived the competition through perseverance, agility and independence."

On the blanket criticism from foreign media about "China's lack of innovation," Wolf had a different view. He called China "an innovative country" and defined Shenzhen as a "Chinese-style Silicon Valley." Wolf pointed out: "China telecom equipment companies pursue what we can call 'customer-centered innovation.' This pursuit leads them on a long path of innovation that keeps on going... Real breakthrough occurs when the customer asks: 'Can you...?' and gets answered not with 'We can' or 'We can't,' but 'Leave it to us!'"

The story of the distributed base station is a Huawei legend. Huawei invested CNY6 billion into 3G (the third-generation mobile communication technology) R&D, but because of China's delay in issuing a 3G license, Huawei's 3G business had almost no income for about three years. Every time Ren Zhengfei met with Huawei's head of wireless products, he would ask, "When are you going to earn that CNY6 billion back?" Bear in mind, back in 2003, only very few developed countries and regions such as Europe, the United States, Japan, Hong Kong, and Saudi Arabia were building 3G networks.

Huawei wireless products penetrated the "iron curtain" in the European market with distributed base stations. Prior to this, customers in Europe had habitual distrust towards the Chinese. In a bid for Telfort, a small operator in the Netherlands, the customer bluntly asked, "Your price is so low, why should I buy from you? You can't guarantee my success." The head of wireless product sales at Huawei was frustrated: "Our price is 20 to 30 percent lower than others, but still got rejected. What exactly are the customer's pain points?" After several exchanges with the customer, he found out their main difficulty was indoor coverage, and said to the customer: "Leave it to us." He drew a few sketches for the customer, and the latter said, "If you can do this, we'll buy it."

In response to the customer demand, Huawei's wireless R&D department gathered a number of its best engineers, who worked overtime for over half a year to come up with a satisfactory product. This led to a breakthrough in the European market for Huawei's 3G products.

The distributed base station is known in the industry as a "disruptive architectural innovation," in parallel with another of Huawei's disruptive innovations: SingleRAN, which uses a complex mathematical algorithm to integrate 2G, 3G, 4G, and future systems together through a single base station. This greatly reduced operators' costs and became a benchmark in the industry. At the same time, it substantially enhanced Huawei's competitiveness in the global market.

Ding Yun, the executive board member in charge of Huawei's entire R&D division, said: "SingleRAN was a revolutionary innovation, but we do not innovate for the sake of innovation. The SingleRAN solution came out of a demand from Vodafone. Huawei invests more than 70,000 R&D staff in technological innovation, but our innovation philosophy is that innovation has to center around customer needs, both visible and invisible."

Li Yingtao, the first director of Huawei's Sweden Research Center, now a board member and the president of Huawei's 2012 Laboratories, repeatedly stressed in a four-hour conversation with the author: "The genes that determine a company's success are the logic behind its choices. Huawei chose to be customer-centric; hence, all acts of innovation that meet customer needs are well grounded. The success rate is much higher this way, because all innovation is centered around customers' visible and invisible needs."

During the period in 2011 when the IT bubble burst, Huawei's current board director Zhang Ping'an actively pursued an opportunity to preside over the acquisition and integration of a Silicon Valley optical transmission company. That company had invested US$140 million in total, but Huawei only spent US$4.2 million on the deal. While competitors believed that optical transmission technology had been overdeveloped and unworthy of further investment, Huawei had keen insight into the company's technology, knowing that it was not only suitable for long-distance transmission, but also great for countries with different optical fiber transmission. In the future, it could also be upgraded to higher rates of transmission, like 40Gbit/s and 100Gbit/s, which would be hugely significant for the development of customer networks.

The use of microwaves in telecommunications is a mature industry that hasn't seen much development for the past 50 years. In 2008, Huawei introduced the concept of the Internet Protocol (IP) microwave for the first time, causing an "arms race" in this industry. The idle transmission rate of traditional microwaves was 2–8Mbit/s; Huawei was able to make it reach 1.6–2.5Gbit/s. This development was also based on customer needs. After base stations are built, customers want to backhaul data received by the base stations at the fastest speed possible and, at the same time, make it easier to configure fixed networks. These two major demands were overlooked for a long time until Huawei came along and integrated IP with microwave technology. Again, this disrupted the entire industry while resolving customers' pain points, which naturally nudged Huawei into the top position in the market. Huawei is ranked number one in optical transmission, core network, and fixed access products globally and number two for its wireless products. Without exception, this is the result of a rigorous focus on customer needs.

Mao Shengjiang, the former president of a submarine cable company, joined Huawei in 1991 and headed the development of the C&C08 2000 program-controlled switches. Reflecting back on that time, he commented: "From early on, Huawei rarely had any detours in terms of technology development. This is mainly because we are very close to our customers—I don't see any telecommunications company around the world that is as close to the customers as we are. In addition, we strongly emphasize the success of our customers, so innovation at Huawei always leads to breakthrough products... Especially in recent years, I myself am amazed by the explosive growth of Huawei's R&D capabilities. It's like a hot knife cutting through butter—whatever we set our minds to, we succeed. IP microwave and SingleRAN were good examples…."

David Wolf pointed out that in Shenzhen, the "big melting pot" where free competition was introduced, Huawei hyped up a Chinese-style innovation, a style that differs from the stereotype of "imitation innovation" in the eyes of Western media.

Renouncing Sentimentality, Clamor, and Glamour

In 2002, the Chinese magazine *CEOCIO* published the article "Evolution from Wolves into Lions". The author divided telecom manufacturers into three types. The first type are lions: Western companies that boast comprehensive advantages in technology, products, capital, and management and that have a self-perceived sense of superiority are typical examples. The second type are leopards, referring to Sino–foreign joint ventures. The third type are wolves, or local companies, such as Huawei, that lack advanced technologies and produce poor-quality products, but are highly aggressive in the market, trying to survive through natural selection. Wolves are serious threats to lions and leopards.

This was the most insightful article on Huawei to date, but Huawei's management did not agree that they were wolves; they even felt insulted.

Huawei refused to be sentimental. Ren Zhengfei once said, "Our company has succeeded because we have not paid attention to ourselves. Instead, we have focused on maximizing value for our customers. It is our goal to create maximum benefits for telecom operators, and we have tried every means to realize this goal." Concerning how the company has fought its way in the market, he said, "Huawei has consistently focused on its customers, rather than its competitors." Therefore, Huawei's culture is quiet and simple. It has two syllables and a single color. There is no redundancy, ambiguity, bustle, or splendor.

While the lions and leopards stopped hunting, cursed as they were with ample resources, Huawei, the wolf, gradually evolved into a lion. How has it managed to do so? Ren Zhengfei explained:

We must adapt to changing circumstances, rather than follow any rigid tenet. The key is to satisfy customer needs. We must act like businesspeople. A scientist can focus throughout his entire life on one hair on the leg of a single spider, but if we do the same, we would starve. Therefore, we must not focus exclusively on spider legs; instead we must study and understand the needs of our customers.

These remarks were made in 2002 when Lucent Technologies, one of the lions, was about to fall and Motorola, another huge lion, was also sick.

The most essential resource of Lucent Technologies, Bell Labs, was known for its research on "spider legs," "butterfly wings," "horse tails," and other fundamental research subjects. Bell Labs had been a boost for the growth of Lucent Technologies, but later became a huge burden. Motorola had invested heavily in its Iridium Satellite system, and this cutting-edge technology threw Motorola headfirst into its own Waterloo. Like most other lions, both Lucent Technologies and Motorola had actually suffered from the resource curse, or the paradox of plenty, and been dragged down by excess capital and technology.

On the other side of the spectrum, Huawei has grown up in hunger. It suffered from inadequate resources, so its aim has been very simple: satisfy customer needs with good products, low prices, and excellent services. When Huawei grew into a lion after becoming the second-largest telecom equipment manufacturer in the world, the company came to possess more ample resources, including capital, technologies, talent, and management expertise. Nevertheless, Huawei's leadership team does not only hold fast to the common sense of customer centricity, but also gives the concept new meanings.

In 2009, Ren Zhengfei visited the Dujiangyan irrigation system, an irrigation facility in the Sichuan Province built in 256 BC. During the tour, he was told the story of Li Bing, the man who planned and built this system, and his son. He was greatly inspired, and later wrote the article "To Dig Deep Channels, and Build Low Weirs," in which he said for the first time, "Competition in the future will be between industry value chains. The robustness of the entire value chain is the key to Huawei's survival."

From 2009 to 2010, Ren Zhengfei continued to elaborate on this idea. In various speeches and essays, he has stated that to dig deep channels means to tap further into the company's inner potential. The company must ensure sufficient investment in its core competitiveness and its future, even during times of financial crisis. From 2005 to 2014, Huawei's R&D investment amounted to CNY190 billion, CNY32 billion in 2013 alone and CNY40.8 billion in 2014, accounting for 12.89 percent and 14.2 percent of its annual sales, respectively.

> "To build low weirs" means the company must not pursue short-term goals at the expense of long-term ones. We have to share the benefits of growth and create more long-term value for our customers. By digging deep channels and building low weirs, our purpose isn't to make as much money as possible, but of course we can't afford to lose money all the time, either. What we need is a modest profit, leaving the rest of the water to overflow the low weirs, reaching our customers and the supply chain. In this way, we will be able to survive. Those who survive the longest are the best, because they have to contend with strong competitors in every partnership they build. Those who survive will become the stuff of legend.

Clearly, this is a significant extension of customer centricity, a core value at Huawei. These statements describe its further commitment to the industry, and also a new mission that Huawei defines for itself: to open up, build partnerships, grow through self-reflection, and be a hero that can accommodate the world.

The 100-Year Business and the 1,000-Year Temple

Supernatural Force: Religion and Values

There was once a very popular, widely quoted short blog post on Weibo, a microblog website: "The Greatest Business Model in the World." In this post, the blogger states that the greatest business model in the world was not created by Steve Jobs, but by the Buddha. Buddhist temples are the most successful chain stores. They don't sell products, but they have the largest number of loyal customers. Buddhist temples are among the most-visited tourist destinations in the world. They have a consistent visual identity, management system, and culture. They don't need to advertise. Customers come to them in droves.

There is another short article with similar content. Titled "Buddhist Temples are Real Estate Businesses Superior to Apple," the article argues that Buddhist temples have quite a few advantages:

1.   They have a clear and distinctive theme and represent universal values and spiritual authority.

2.   They have a huge number of loyal believers. People from all walks of life—from beggars to billionaires to government officials— have strong faith in Buddhism.

3.   They have a unique profit-making model. There is no compulsory consumption, yet they still make a lot of money.

4.   They can satisfy huge demand with very few resources. They are typically located in remote areas, but they still have many followers.

5.   They operate nationwide and pay no taxes, yet enjoy government support wherever they are.

Although these articles may seem frivolous or even profane, they both touch on quite a few questions that may point to the truth: Why does religion possess such timeless power in communication? How can religious faiths like Christianity, Islam, and Buddhism last thousands of years? How have they crossed the boundaries of nations and of socioeconomic class?

What is the answer to these questions?

As Plato argued, Forms are the most real existence in the world. The grandeur and endurance of religion is in part because it represents a supernatural force that exists in the superhuman world of Forms. Religion is directly attuned to our hearts, transforming humanity and guiding our ethics and behavior. More importantly, in the wisest possible way, it answers the ultimate questions we have as human beings: Who am I? Where do I come from? Where am I going?

Religion then extends to the concepts of redemption and universal salvation. It allows people to understand what it means to exist and provides them with spiritual support. To put it metaphorically, religion points out where the river is flowing and at the same time prepares the boat with which to cross the river.

Buddhists believe in the law of karma, with strong punitive and incentive implications: If you practice goodness in this life, you will be blessed in this life and the next; if you are a proponent of evil, you will be punished in this life and enter hell in the next.

There are many rituals in Buddhism, such as praying, scripture recitation, incense burning, and meditation. A Buddhist temple is also characterized by grand and solemn buildings and slow, pacifying music. These are all physical vehicles of the religion that signify, demonstrate, and pass on its values.

Will religion one day fade away? According to the Buddha, everything in the world has a life and every life will eventually end. This is also true with any religion. Religious belief, however, is greater than any human organization because it can adapt to the vagaries of time and circumstance. It finds new ways into our hearts through evolving on its own, while still maintaining its core values.

It's rare that any political, governmental, or business organization can achieve this.

Common Sense, Pushed to the Extreme, Is Religion

To some extent, the most outstanding business leaders in history can be compared to religious leaders. In the first instance, these leaders practically worship the perfection of religious organizations, and such worship is intuitive and instinctive. As a successful businessman, Steve Jobs was also an artist, a dictator, a priest in the black gown, and a godfather who had stubbornly worked for extreme perfection all his life.

During his last years and even after his death, he had billions of fervent followers who remained loyal to him. He was their godfather or guru. *Time* magazine listed Steve Jobs as the most celebrated, successful business executive of his generation, stating that he will be remembered for at least a century, for he is far greater than any of his contemporaries.

In what ways was Steve Jobs so great? *Forbes* answered: Jobs knew what people wanted. James Marshall Crotty writes:

> Early Jobs was satisfied with producing innovative, beautiful, yet deceptively simple hardware and software. Late Jobs wanted users to feel welcomed and heard. The Apple Store is the clean, bright, and ordered example of that ethic. In my view, it is Jobs' most enduring legacy… The Apple Store is the human glue that connects the user with technology in a way that is precise, compassionate, and patient. Call Mr Jobs what you want, but his greatest achievement is that, in the end, he understood what all us in the 21st century secretly need to hear: a friendly voice reassuring us that machines are not in charge, we are.[2]

Jobs was considered the consummate genius of the business world because, to him, the customer's needs were the be-all and end-all of his business. He was born with a sensitive heart. He knew that truth could be found in common sense and that common sense pushed to the extreme was religion. This process of transforming ordinary consumers into product addicts, and then to loyal followers of the product designer, is not unlike the story of creation itself. People remain loyal to God not only because God created them, but also because God has taken good care of them.

So therein lies the question: How can religion last so long, while businesses die so soon? This is a question that troubles almost every entrepreneur. To build even a century-old business has already seemed an elusive goal. Microsoft, for example, was once hailed as the best in history, but it remained on top of the world for a mere 50 years. Now the sun is

---

[2] James Marshall Crotty, "Steve Jobs Reinvented What It Means to be Human," *Forbes.com.* Forbes, Inc., October 6, 2011.

beginning to set on this particular empire. Similarly, we may also wonder how long Apple will last now, in light of Steve Jobs' death.

On a sunny afternoon, I visited Wenfeng Temple in Lijiang, Yunnan Province, and had a talk with Tian Liang, a secular disciple of Dongbao Zhongba Rinpoche, the living Buddha. Tian Liang was about 30 years old at the time. He had worked for Hunan Television as an editor and director, and was once a keen paraglider and mountaineer. He was acquainted with Wang Shi, the former chairman of Vanke Real Estate, and other entrepreneurs. A year ago, Tian withdrew from all secular activities. He came to Lijiang and became an informal lama, or a secular monk. Tian said:

> Isn't the culture of a company like a religion? You need to motivate every employee. Money is not enough. You need to give them some beliefs and values that will inspire them to work hard. You need to do the same with your customers. If we compare employees to monks in a temple, customers are believers who come to worship the Buddha and pay their respects to the temple. If you want them to buy your products, you need to care for their needs.

He uses an Apple phone, and I bet he knows about Steve Jobs.

On a flight from Lijiang to Beijing, I sat in the first row, next to Dongbao Zhongba Rinpoche, the *tulku*. After a simple meal, many passengers were waiting to use the restroom. The living Buddha said to me:

> Everything changes; this is universal truth. A delicious meal turns into foul-smelling waste a few hours later. This is a natural change. You may be glorious now, but sometime later the glory will go. If you want to be blessed for long, you need to practice discipline for that length of time.

This is as true of business organizations as of people.

All Methods Are a Practical Means to an End

Things change; no company will last forever. For any entrepreneur, it is a dream come true to have a company that lasts a century. To that end, business leaders and management experts have engaged in various studies and practices, proposing one theory after another. They dig deep into human nature and the concept of product cycle; they apply innovative business models, strategic management, performance appraisals, and team development; they also innovate in management and technology. All such endeavors are useful, but they are only a practical means to an end. A fundamental Tao—a universal principle, or path—underlies them all.

What is the Tao of business? "Customer centricity." This is a piece of common sense that can stand the test of time and circumstance. Religions are able to last so long because they are built on common sense. Steve Jobs' combination of simplicity and aesthetics exemplifies the notion that success is based on common sense.

Ren Zhengfei once said, "We will always treat our customers with religious faith." He added, "Serving customers is the only reason Huawei exists. This should be a true belief of our people, and cannot remain a mere slogan. We need all our employees to take action. Huawei has only one clear value proposition: serve our customers."

Ren is an adventurer. No one can become an entrepreneur without a risk-taking disposition. Entrepreneurs are sailors navigating the sea or knights riding through the wilderness. Generally speaking, Ren is not an extremist, but he has been stubborn to a fault when it comes to company values: "One cannot succeed without insane stubbornness." This is a perfect statement to describe Ren's position on corporate values. He is neither Christian nor Buddhist, but he uses the word "religious faith" in connection with customers. Time and again, he has emphasized the value proposition of customer centricity with uncompromising modifiers, such as "the only way" and "always."

In our times, capital and technology are worshiped. Huawei is an exception. It worships a god, but that god is not of the Christian or Buddhist traditions. Huawei's god is its customers: over 700 telecom operators and approximately one-third of the world's population.

Huawei now employs 170,000 people, most of whom are knowledge workers. Before joining Huawei, each had his or her own unique personality and a different set of dreams. Many were romanticists or idealists. However, after they entered the company, they were regularly instilled with the company's core values. Having gone through various training sessions on the values that drive their daily work, almost every one of them has undergone a transformation: Every cell in the Huawei organism is trained to be customer-oriented. The people, workflows, business processes, R&D activities, products, and corporate culture at Huawei are all infused with life and continue to live by the grace of customer service alone, without which they would perish. At Huawei, the truth takes the place of fantasy, implementation is more important than creation, and performance is more valuable than the process. Not a single person or a single thing can deviate from their focus on the customer.

Common sense, pushed to the extreme, is religion. This is true for Apple, and it is also true for Huawei.

For over two decades, Huawei has never swayed from its core values. Even after Huawei became a world-class company, Ren Zhengfei remains vigilant. He has the following to say:

> To better serve our customers, we should locate our command posts in places that are within earshot of the gunfire. We should delegate planning, budgeting, and accounting rights, as well as the right to make sales decisions, to the frontline, and let those who can hear the gunfire make the decisions. The back office can decide whether we should engage in a battle, while the frontline decides how to fight the battle. The back office should follow the instructions of the frontline, not vice versa. The headquarters is the support, service, and supervision center; it is not the command center.
>
> Who shall call for artillery? That decision must be made by those who can hear the gunfire.
>
> We have established an "Iron Triangle" in our account department to identify and seize opportunities, move operational planning out to the field, and to summon and organize the forces necessary to hit our targets. The Iron Triangle is not for checks and balances. Instead, it is a customer-centric joint operating unit where different roles are closely connected, and cooperate towards a common goal: meeting customer needs and helping customers to realize their dreams.

The Soul of Business: Dedication Is the Key to Success

Chapter 2

A Fortune 500 Company That Rolled in on a Pushcart

Where Huawei Made Its Very First Pot of Gold

The first time Guo Ping, currently one of Huawei's three rotating CEOs, met Ren Zhengfei was in 1988. He was then a postgraduate student at Huazhong University of Science and Technology. Ren had embarked on a long train ride from Shenzhen to Wuhan, carrying a 40-line PBX, to come visit Guo's supervisor, who was one of China's very few experts in PBX technology. Wu Jiangxing, known in China as the "father of large-capacity PBXs", was once mentored by Guo's supervisor as well.

In those days, Huawei was a trader of switching equipment, a so-called "dirty-dealing reseller" in the sociopolitical climate at the time. It would buy PBX systems from a company called Hongnian in Hong Kong and another company in Zhuhai, and sell them to county-level post offices, small towns, and mines. Hundreds of companies with a similar business model existed in China at the time, enjoying extremely high, very comfortable intermediary margins. The people who ran these companies were among the first in China to grow wealthy, and their companies were known as "teabag companies." This was because the owners of these companies were all noted for enjoying high tea in the morning and acting like teabags themselves in the evening, soaking in the pool at the spa.

Where did Huawei come across its first pot of gold? Through a model of reselling at a profit and credit-based transactions, the company would first take on goods, then settle their account with Hongnian after the goods were sold. Huawei's accounts payable to Hongnian were in effect a zero-interest loan of over CNY100 million. The only problem was that this supplier was often out of supply. Many old-timers at Huawei share a fond memory: Whenever someone yelled "The supply arrived!" outside the office, the entire company, including Ren Zhengfei, would cheer loudly and rush downstairs to unload the goods. "It felt like a New Year's celebration," someone recalled. If it weren't for Ren's grand dream and the ambition of the people around him, perhaps Huawei would have drowned in the tides of history too.

Destiny isn't a predictable thing. But when you take a turn at its wheel, certain choices became more probable than not. When he served in the Infrastructure Engineering Corps as a deputy regimental-level technology officer, Ren was the mastermind behind two inventions that filled a technological gap. If his daughter Meng Wanzhou had not complained, he might have stayed in the mountains in Sichuan and became a scientist. And if he had not been transferred to Shenzhen, a "cultural desert," he is unlikely to have become a businessman. Deep inside, he has a strong psychological connection to technology. "I can't control the market," he says, "but I have to find something that I can get a grip on. I think that something is R&D, developing my own products." Huawei's earliest registered business scope was the development of gas suspension instruments, which were Ren's own invention.

Ren later chose to focus on the telecom industry. History shows that this was certainly a wise choice, but the path he chose was bumpy.

History is a story of completion in itself. No matter from which angle you scrutinize Huawei, it presents a rather pedestrian logic: Huawei's rise to prominence is through collective dedication, creation, resilience, and the cohesion of its teams, not luck or any other mysterious external factors.

Back in the 1990s, the early entrepreneurs who had founded their own reselling outfits at around the same time as Ren used to show off their wealth by driving imported Japanese cars, like Toyota Crowns or Nissan Bluebirds. Huawei, however, started out "climbing the Himalayas from the north side," according to Liu Chuangzhi, the founder of Lenovo.

Only those who fight are able to survive. Among all the different types of management, industrial management is relatively difficult and complex, and of all industries, management in the electronics industry is hardest of all. Compared to traditional industries, the electronics sector features relatively fast-paced technology replacement and industry-wide changes. At the same time, it's not as restricted by natural factors as, for example, the automobile industry is restricted by steel, petroleum resources, and road infrastructure. The electronics industry draws its raw materials from an inexhaustible supply of river sand, software code, and mathematical logic. The nature of this industry is what led to fierce competition and more ruthless elimination than in traditional industries. Lagging behind is tantamount to biting the dust.

In 1991, Huawei burned its boats by deciding to concentrate all of its capital and human resources into developing and producing Huawei's own model of PBX product. They rented three-storey space in an industrial building and sectioned it off into four workstations, including a warehouse and a kitchen, and set up dozens of beds against the walls. There were not enough places to sleep, however. Mattresses placed on foam boards were also used as makeshift beds. Some 50 R&D people lived in this space, working together day and night with the company's managers. At one point, an engineer had been working so hard that his cornea fell off and he had to be hospitalized to save his vision.

In the end of December that year, their equipment tested successfully, and Huawei at last had its own product. By that time, Huawei's balance sheet was close to zero—bankruptcy was right around the corner if the company wasn't able to deliver its product.

This was a make-or-break battle, and Huawei won. The founders of the company realized a profound truth: There is no shortcut to survival and development; despite how stupid or elementary it sounds, fighting hard and persevering is the only way.

Huawei came face-to-face with innumerable crises in the history of its development—one executive referred to them as the "81 great trials and tribulations"[1]—and yet, miraculously, they always managed to pull through.

A Fortune 500 Company Pulled along by a Donkey

In 2014, Huawei was number 285 on the Fortune 500. Wang Haitun, responsible for project delivery, commented, "This is a Fortune 500 company that rolled in on a pushcart—that got pulled along by a donkey!"

Huawei is the world's largest telecom engineering contractor and systems integrator, possessing a powerful arsenal of what it calls TK (or turnkey) solutions, which have helped secure its success in the global market.

So what are TK solutions? A TK solution means that Huawei's customers are only responsible for investment and spectrum acquisition, while Huawei is responsible for the whole package deal, from network design, topography survey, site acquisition, pipeline setup, equipment transportation, and network testing, to full system readiness—essentially everything up to final approval of delivery. Today Huawei works with about 500 operators in 170 countries and regions. Many of these projects were won with TK solutions.

Sweat, tears, blood, life-threatening danger, and countless unimaginable hardships all go into each TK solution. Take site acquisition, for example. In many remote areas of Africa, South-East Asia, and Latin America, the Huawei staff has had to negotiate with local chiefs, warlords, non-governmental armed militia, village tyrants, and the like. In the Egyptian–Israeli land bridge, the Sinai Peninsula, they had to negotiate with local

---

[1] "81 Trials and Tribulations" is a reference to the famous Chinese masterpiece *Journey to the West*. The story features a Buddhist monk named Tang Seng, who was instructed by the Buddha to set out and obtain the sacred texts of Western religions and then bring them back to the East. Together with a company of mythological misfits, his journey takes him through mysterious lands and 81 different perils, trials, and tribulations. Here, this phrase is used to indicate every possible difficulty—the Chinese equivalent of the Twelve Labors of Hercules.

military heads, and in the Suez Canal, their customer was quite explicit about being unable to help with site acquisition and told Huawei staff to figure it out on their own.

Sometimes, dozens of tons of transmission tower parts were air- or sea-transported from China to locations thousands of miles away, and were trucked out to where the roads had ended, only to find gullies, muddy swamps, steep slopes, or an illimitable desert ahead. In such circumstances, their only choice was to adopt the most primitive methods available—they employed local villagers, pushcarts, and donkeys to transport the equipment to tower sites inch by inch, taking hours or even days, with Huawei engineers (often one or two people) trudging along with them. The engineers would begin installation after they arrived at the site and sometimes, when the equipment didn't fit, they would have to come up with solutions on the spot. Most of these areas did not have heavy duty cranes and other lifting equipment, so installation was mostly done manually, often taking 10 or more hours. After installation, they would have to hire guards to prevent the theft of power supply equipment.

In remote, undeveloped areas, it would normally take an average of more than three days—sometimes more than 10—to transport and install a base-station tower. So it was that Huawei's people covered more than half of the world at a snail's pace, traveling from one pole to the other, from the Himalayas to the Sahara desert.

When face-to-face with customers, Huawei seldom says no; the answer is always yes. Frontline employees would often send back a pile of impossible contracts to Huawei's R&D, delivery, supply-chain, and service functions at HQ. On the one hand, these contracts led to a number of customer complaints. But on the other hand, in most cases, they served as the catalyst of countless miraculous deliveries. An executive who joined Huawei from a Western company observed: "For a very long time, Huawei's management was utter chaos. People within and across departments were constantly at battle with one another; there was no method to the madness. However, the company had a spirit that Western companies lacked: When someone stood up and made a promise to the customer, a bunch of fearless fellows would line up, determined to fulfill that promise no matter what—whether they were capable of doing so or not—and often they'd succeed. As the saying goes, 'throwing your fists around like mad will kill the Kong Fu master.' And this is where Western companies got whopped."

Wang Haitun, director of delivery, described his own experience like this:

> When it comes to delivery, we have to aim for the stars. If we can hold on to the thought 'Where there is a will, there is a way,' then we are likely to succeed. Back in 2006, Huawei signed a US$700 million exclusive, all-network system contract with Ufone Pakistan. Such projects normally take three years, but the client requested delivery in 12 months. In the second half of the same year, Brazil's largest telecom operator Vivo signed a US$60 million contract with us, requesting that we complete the transfer of their network from CDMA (code division multiple access) to GSM (global system for mobile communications) within three months. In early 2007, Huawei's rep office in Egypt promised a customer we would deliver a US$170 million TK project that involved over 1,000 base stations. The contract was not finalized, so the rep office had to take loans from the head office. The customer had capital and spectrum resources, but no telecom operations experience at all, and they wanted Huawei to deliver within six months....

One after another, three massive projects all landed on Wang Haitun's desk. Aside from strengthening internal coordination to streamline R&D, production, and logistics, and proactively communicating with their customers, they only had one other way to meet their deadlines: working overtime from 8 AM to 1 AM every day, giving up all holidays and vacations, all the way to up to final delivery.

This was how things were in delivery, and other departments were no exception, including marketing, R&D, logistics, and so on. An employee once commented with bittersweet pride that Huawei's people spent an exhausting 27 years of their youth to defeat century-old competitors, one after another after another.

The Soul of Business: Ren Zhengfei on the Culture of Dedication

The Goal of Dedication: Customer Centricity

In 1994, Li Jie, who had joined Huawei less than two years prior, was transferred to supervise sales and marketing. Ren Zhengfei asked him in a meeting, "How many counties can you cover in a year?" Li gave an impulsive answer: "Five hundred!"

"Five hundred it is then. Go get 'em!" Ren replied.

Just like that, Li, along with a group of a dozen people, drove half a dozen Mitsubishi Jeeps and two Audis from Shenzhen to post offices in different counties across China, promoting Huawei's newly developed central office switches. Each one of them covered 40–50 counties, and each county took about three days. In less than two years' time, they ended up covering 500 counties, and accumulated a pile of customer files a few feet thick. This was so unheard of in the entire history of telecoms that the Minister of Posts and Telecommunications at that time, Wu Jichuan, told his own departments in a ministry meeting that they needed to learn from Huawei.

In August 2007, Huawei was contracted to build two mobile-communication base stations for China Mobile on Mount Qomolangma (Mount Everest) at different altitudes: 5,200m and 6,500m. According to the contract, the base stations needed to be deployed by the end of November 2007, but the circumstances were extremely challenging. At those heights, people can hardly do anything because of the mercurial weather and the extreme thinness of the air. Oxygen levels at an altitude of 5,200m are 50 percent lower than those at sea level, and they were even thinner at 6,500m. Nevertheless, four engineers from Huawei, together with one driver, made it up to the rooftop of the world with protective clothing and footgear, in addition to special mountaineering equipment and food that was made for harsh environments. Dizziness, headaches, swollen lips, ulcers, and sleeplessness were commonplace, and one of them suffered nosebleeds for two days. When camping at 6,500m, they would awaken in the middle of the night to find their hair had frozen.

Huawei completed its 3002E base stations at 1 PM on November 13, 2007. Since then, all camps and mountaineering routes on Qomolangma have had access to a mobile communications network. Huawei had built the highest wireless-communication base station in the world.

On May 21, 2003, Algeria was hit by an earthquake of magnitude 6.8, killing more than 3,000 people. Soon after the earthquake, all of the expatriate staff working for Western companies had left the country. The Huawei staff however decided to stay. Three days after the quake, Huawei's engineering team completed an intelligent-network cutover as scheduled, which ensured local access to communications networks during a critical period of emergency response.

"The most painful memory was not the earthquake itself," describes Chilean author Alberto Fuguet, recalling the 2010 Chile earthquake, "but the aftermath distress where your phone stopped working and you couldn't get through to your loved ones." What he described was also felt by the 33-year-old Huawei employee Sun Dawei and two local employees, Molina and Perez. During this earthquake of magnitude 8.8, the three of them went against the flow of people escaping the disaster area, carrying with them diesel fuel, food, and water.

"Fear and restlessness filled my heart, as if a huge black hole was waiting for me." But, Sun continued, "Every person at Huawei knows that actions speak louder than words, and we do this so our customers know that we're a trustworthy partner." In order to meet customer needs, the three of them stayed in a hotel with cracked walls and split floors, brushed their teeth with swimming-pool water, and ate only bread for five days, working together with the client day and night to repair the failed equipment until the site was up and running again.

Then there was Mumbai, India. One day, the city was struck by a terrorist explosion. The streets emptied and all shops closed down. There were barely any vehicles on the road. In the early hours after midnight, despite the danger, Huawei's engineers rushed to the customer's site and upgraded their communications facility to meet a 5:30 AM deadline.

The earthquake that hit Japan on March 11, 2011, triggered a massive tsunami and aftershock that reached as far as Tokyo. Western companies all evacuated at the earliest possible moment. However, Yan Lida, then head of Huawei Japan (currently president of Huawei's Enterprise Business Group), is a firm believer that "especially in critical moments, we need to fulfill our corporate social responsibility." He wrote a long English email to all employees, explaining the rationale behind the company's decision to stay and describing relevant safety precautions. As a result, a scene that deeply impressed many local Japanese unfolded before their eyes: While tens of thousands of people evacuated the epicenter of the disaster, Huawei's employees—both Chinese and local Japanese—marched into the disaster zone, luggage on their backs. In the meantime, Wang Shengli, Huawei's Asia-Pacific president, and Chairwoman Sun Yafang also rushed to Japan to visit the staff and communicate with customers amid the intermittent aftershocks.

During the Wenchuan earthquake in Sichuan, China, Huawei engineers guarded the temporary installation of communications equipment for several days on the top of a shaking hill. The company hired counselors from Taiwan to connect with the team and comfort them every few hours.

A Huawei executive summarized:

> What is customer centricity? It is not about bowing and scraping to customers; it means accountability to the network and loyalty to our job. We have to start with a full understanding of customer requirements and needs, and do everything we can to meet them. When our customers choose to build their networks with our equipment, we must make timely, accurate, high-quality, and low-cost delivery, and provide the best service. Under extreme circumstances, such as earthquakes and civil unrest, we must stay and assist our customers through the difficulty because the network is most vulnerable at such times.

Customer centricity is the basis of dedication. Ren Zhengfei said the following:

> What is dedication? Someone is dedicated if they create value for our customers, no matter how insignificant that value might be.

You are dedicated if you try to improve and advance yourself in order to better serve customers; otherwise, you cannot call it dedication, no matter how hard you work.

In 2010, a senior executive elaborated on the same idea:

> We believe in customer centricity and believe that dedication is the key to our success. This belief has been distilled for over 20 years, and is core to our corporate culture. We are committed to providing timely, accurate, high-quality, and low-cost services to our customers, and this is the aim of everything we do. Customer centricity is common sense. Without our customers, we would have starved. Inspiring dedication among our employees is, in essence, also being customer-centric. If whoever serves our customers well is considered the backbone of Huawei, and if they share in the benefits of the company's growth, then we are fostering customer intimacy and a customer-philic force within the company. Long-term dedication is also customer centricity. Every cent we spend comes from the pockets of our customers, and they won't allow us to spend their money for nothing. If you are afraid of working in a difficult environment or performing a difficult job, if you don't focus on our customers, our customers won't accept it, and they won't recognize your value. That's a hard life to live.

Mattress Culture

In the early years of the company, every new employee was given a blanket and a mattress. During their lunch break at noon, they would take a nap on them—both convenient and practical. When working late into the night, many of them would rather stay in the office and sleep on their mattresses instead of returning to the dormitory. It was convenient because they could get up at midnight and continue to work. As many employees said, "These mattresses meant a great deal to us. They represented relentless hard work in the old days, and now this habit has been incorporated into our concept of dedication. They offer a unique glimpse into Huawei."

Zhang Yunfei, whom Ren Zhengfei called a "software guru," has worked in Huawei for more than seven years now: He is in charge of software development. In his first few years, he worked and slept in the office almost every day. A dozen mattresses were lined up against the walls of a large office for people to rest on. No one was required to clock in or out, but everyone worked overtime till deep into the night. When everyone went to sleep, Zhang would check their revised code and then integrate it into a new version. He would load the new code on a machine and test it, releasing the code if everything went smoothly. By the time he finished all of this and went to sleep, it would usually be dawn. This sort of inverted sleep routine, held over a long period of time, ended up giving Zhang a severe case of insomnia.

Any fight—any struggle—always comes at a price. Most of Huawei's founding members and senior executives have gone through long periods where they couldn't go to sleep without sleeping pills.

In 2008, a few Huawei employees committed suicide. These tragedies triggered a massive thunderstorm in the domestic media, which collectively lashed out at the company's "mattress culture" and concept of dedication. One Huawei executive explained:

> When we started the business, we had only five or six engineers in the R&D team. They had no resources, and little to rely on but their own two hands. They drew inspiration from the tough and unwavering spirit of Chinese scientists who devoted their lives to developing China's first atomic bombs and satellites in the 1960s. With a completely selfless devotion to their work, this old generation of scientific workers are the best role models out there. Our own engineers worked day and night on their R&D projects, developing, testing, and validating our products. They didn't take holidays, and even gave up their weekends. For them, there was no difference between day and night; whenever they felt the need to rest, they would take a nap on the floor and would wake up rolling their sleeves again. This is how the "mattress culture" got started. We now use mattresses mostly for naps at noontime. In some sense, this bedding represents the hard work of Huawei's early generation of pioneers. This is their spiritual legacy, a legacy we highly value across the company.

Two years later at the company's annual market conference, Ren Zhengfei said, somewhat indignantly:

> Some people are criticizing our culture of dedication! I say, what's wrong with fighting to achieve something? We can all learn something from the tradition of dedication in the Communist Party of China (CPC). 'A lifetime without rest, struggle that never ends,' all to make the country better—these are the mottos of the CPC, and our sense of dedication is no different."

In response to some scholars who urged China to leave behind the concept of "Made in China," adopting instead a "Created in China" approach, Ren said:

> People tend to neglect or ignore the fact that creativity grows over a long period of time, and at terrible cost. Many companies have to die off before a handful of others can emerge and shine. We have gone through purgatory over the past few decades, and only we ourselves and our families can possibly know how tough these years have been. We have not worked 40-hour weeks. In our early years, I worked 16 hours a day. I had no house of my own; I had to live and work in the same rented apartment. I had no weekends or holidays. I wonder if you can imagine that all of our employees, over 100,000 of them, have dedicated themselves to the company over the past 20 years. This is not only true of our current employees, but our past employees as well. Industrial transformation is no easy job; it cannot be accomplished in a short amount of time, or by working only 40 hours a week. Forty-hour weeks will only get you ordinary workers, but you'll never get musicians, dancers, scientists, engineers, or businessmen.

When asked by an American entrepreneur how his company became a world leader in 30 years, the CEO of a famous Chinese company once said: "Actually, I consider it 60 years, because we worked two times 8 hours every day..."

There is another question: Would dedication to a company, or any hard work, necessarily lead to suicide? The answer is not that simple. Working conditions in Huawei were extremely poor during its early years, but people were filled with idealism, passion, and entrepreneurial spirit.

No one committed suicide. In recent years, however, there have been reports of students or teachers committing suicide, even at prestigious universities. According to an essay published in *The Lancet*, the world's leading general medical journal, the annual suicide rate in China is 23 per 100,000 people, which is 2.3 times the global average of 10 per 100,000. The number of suicides per year in China accounts for one-third of the world's total. There are now thousands of suicide-related websites on the Internet, and at one time, the Doomsday 2012 prediction had spread throughout the world.

Have we perhaps entered an age where pessimism pervades the collective unconscious? Against this backdrop, managing a company is bound to become more and more difficult.

'Wolf Culture': Vision, Will, and Character

"He who can endure what others can't will be the one who leads the pack." This is a universally acknowledged logic that applies equally well to personal growth and company development. The people at Huawei have always believed that there is no such thing as free lunch. In 1994, Huawei participated in the PT/Expo Comm China exhibition for the first time, where it had this message in its pavilion: "We shouldn't count on saviors, immortals, or emperors. We have to create a new life with our own two hands." This message is an apt description of the past and the present of the company.

Since 1992, Huawei had invested CNY1.6 billion in developing its GSM system and was able to obtain the network access license for it as early as 1998. But the Chinese market was monopolized by Western companies like Motorola and Ericsson, and as a result, Huawei could only secure a very limited share of the wireless market in some small suburban areas. After eight years, the company couldn't even cover its costs in the domestic market. Huawei was forced to embark on an uncertain journey of expansion in the global market.

As the biggest emerging market in the world, China has attracted almost all of the huge international companies out there, throwing Huawei into the most intense competition on its home turf since the day the company was founded. It had to fight for survival in a narrow space. By the time it began to explore the international market, all the fertile land was occupied by Western companies. The only opportunities remaining were in remote areas, countries or regions in turbulence, and places with harsh natural conditions where Western companies were hesitant to make larger investments. To seize this last bit of hope, countless Huawei employees left their families behind to work overseas.

Huawei has been moving forward in this narrow space ever since. It had no choice but to fight for survival in this seemingly impossible war. The key was resilience and die-hard persistence. The company was in such a hard-pressed environment that the so-called wolf culture began to take shape.

In a 1997 article titled "Developing Organizational Mechanisms Suitable for the Company to Survive and Grow," Ren Zhengfei cited for the first time the concept of wolves, saying that the organization should create a structure and an environment where wolves could thrive and maintain their aggressiveness. He said that business managers with strong drive should be developed and motivated to be as tenacious and united as wolves, while the company worked to expand its market. They should be able to sense and seize every opportunity to expand the company's product portfolio and increase its market presence. At the same time, the organization should foster a group of Bei[2], or officials who could develop the right management infrastructure to harmonize internal operations and provide the frontline wolves with the support they need.

The following year, Ren Zhengfei wrote another article titled "How Long Will Huawei Hold Its Banner?" which proposed for the first time that the company should develop a "wolf culture." He said:

> A company needs to develop a bunch of wolves. A wolf has three features: a sharp nose, tenaciousness, and a pack mentality. These are essential for a business to expand. Therefore, we need

---

[2] Mentioned in Chapter 1, the creatures in Chinese mythology that are almost exactly like wolves, but with extremely short front legs and long back legs.

to offer a relatively flexible environment in which people are motivated to fight for success and to hunt down new opportunities that may arise. The sales department has developed an organizational plan that aims to balance the aggressiveness of wolves with sound administration of the Bei. Of course, this plan applies to the departments responsible for business expansion. Other departments may select managers that reflect their own goals and missions.

Huawei is a strong advocate of "the wolf spirit," but Ren's concept of the wolf spirit has nothing to do with cruelty or inhumanity, but instead draws from wolves' acute sense for finding opportunities, their strategic vision, their strong drive to fight and forge ahead, and their esprit de corps—in short: vision, will, and character. Aren't these three qualities advocated unanimously by various human civilizations?

As was written in the press article "Evolution from Wolves into Lions,"

> What are wolves in the eyes of a lion? They attack the lion in a pack of 100 to 1 and eat into its territory from the very periphery to the core. They offer unbelievable prices until the profitability of the lion plummets. The strategy of wolves is to use various unconventional tools and navigate through complex relationships, depending on their unmatched adaptability and market understanding to render the technical advantages of the lion irrelevant.

This is a vivid picture of Huawei 10 years ago. It may be incomplete or overstated, but it largely portrays the organization as it was back then. As of 2002, Huawei has its own set of first-class products and technologies, and has withdrawn from the price war. Yet another picture painted in that article presents Huawei in a new light:

> Wolves are eager to win, but they are extremely resilient in the face of any resistance or frustration. They adapt themselves perfectly to the environment, and they tend to fight for survival in packs regardless of price. In a word, they are fierce and tough threats to lions.

Huawei is not just a local wolf; it has become a "giant wolf" in the global market. To a large extent, it has evolved into a lion itself. What preoccupies the leaders of Huawei is that, once the company becomes a lion, the spirit of the wolf and the spirit of dedication may get left behind. They know very well that there are pitfalls to success and prosperity. At any time, what drag down a company most are surely the company itself and its leaders.

Learning from Its Neighbors: Huawei Won't Die of Complacency

Clocks with Static Pendulums in Rome

One day I was hosting Zeinal Bava, the CEO of Portugal Telecom, and his colleagues in Beijing, where we were having dinner at the Chang An Club. During dinner, Bava mentioned Huawei and Ren Zhengfei. He said Ren had once asked him a question: "At one point in history, Portugal owned practically half the world. Why had it since confined itself to a small corner in Europe?" This question shocked Bava and his colleagues. Yes, in the 16th century, Portuguese adventurers and navigators had hoisted their flags in nearly every part of the world, and a large proportion of the world's territory belonged to Portugal, a small country of several million people. Most Portuguese stopped adventuring after they had come to enjoy a luxurious life with the gold and silver they had plundered from Latin America. Both rulers and ordinary people of the empire had turned speculative and lazy. Navigation and industry withered, eventually dying out. The wealth that had flowed in like water flowed away just as quickly.

Italian clocks and watches are famous around the world. In public places in Venice, Florence, and particularly Rome, clocks are seen everywhere—in shops, hotels, cafés, churches, and street squares. They look exquisite and elegant. But people are often surprised to find that those clocks are not precise and that some have pendulums that have stopped altogether. They are mere decorations, it turns out: Time is immensely abundant for Italians. But the question is, in the past, had Italians really taken time so lightly? If you visit the Colosseum built 2,000 years ago, the churches and museums that have been cherished for centuries, the mercantile city of Venice, and if you see the water taps in Rome that have been used for over 2,000 years, you will be amazed by the achievements of this country. You certainly wouldn't think that they could have achieved so much if they viewed time in such a frivolous way.

On an October afternoon in 2014, I attended a lecture on "Innovation in Switzerland" by IMD (International Institute for Management Development) professor Georges Haous and later exchanged some thoughts with him. Haous said that European intellectuals found themselves in a common state of anxiety and unrest. Criticism was popular, and anxiety had boosted the development of the psychiatric industry. I asked why, and his answer was that Europe had lost its enthusiasm compared to the United States and China, whereas China was still in the midst of an enterprising craze.

As early as in 2003, Ren Zhengfei predicted that Europe would be ruined in the end by its welfare culture. This thought also gave him pause: Would Huawei fall victim to its comfort as well?

In the past 30 years, mankind has experienced globalization at a scale yet unmatched in history. However, in this wave of globalization that was designed and regulated by elite stakeholders in the West, who benefitted the most? Undoubtedly, it was the developing countries represented by China. Why did the US government and some US companies that used to advocate trade freedom start engaging in trade protectionism against Chinese companies like Huawei? At the World Economic Forum in Davos a few years back, John Chambers, the CEO of Cisco, told Ren Zhengfei without an ounce of reservation: "America is my backyard." The fundamental reason behind such a comment was that Chinese enterprises like Huawei practically flew into leading positions in the global value chain within 30 years' time, posing a huge threat to the so-called "international division of labor" theory.

How did companies like Huawei manage to go global, from humble roots in a developing country to achieving a dominant position worldwide? First, the Internet introduced a wide range of information-sharing capabilities, resulting in global mobility of technology, capital, talent, and managerial knowledge.

Second, Western countries largely got caught up in the "civilization trap," relying on huge amounts of debt to maintain a high level of social welfare. Ever since the 15th century, medieval religious reform has liberated mankind and expanded on existing concepts of human desire, promoting a spirit of adventure, sacrifice, and passion for struggle and expansion that was embodied by the likes of the Columbus and his peers. Despite wars, plagues, and natural disasters, the Western world has managed to reach the absolute height of civilization—both materially and spiritually. However, 60 years of prosperity and peace after World War II led many Western countries into a state of arrogance and complacency—marked by uncontrolled consumption, collective inertia, and an overdependence on welfare, especially in Europe.

Third, and perhaps most importantly, the material and spiritual hunger stirring within more than one billion Chinese people were huge drivers for social activity and the large-scale pursuit of development; thus China marched forward with the same voracious appetite that its Western counterparts had a century prior. Hundreds of millions of Chinese people took part in the largest act of urbanization in human history. Meanwhile, tens of millions went out into the world, hungry for wealth, knowledge, and personal success, some harboring a sense of mission for their country and nation—21st-century equivalents of Columbus, but from the East. Huawei and its people are the very embodiment of this latter group.

The key to the rise and fall of any people, country, or company, lies in the vitality of the individuals and the groups within. Vitality is the soul of an organization, while inertia, by contrast, is like cancer.

170,000 Solitary Fighters Marching around the World

The United States is an exception among Western countries. Although it was hit by the bursting of the IT bubble at the turn of the century and the financial crisis in 2008, there is no sign that the United States has entered a state of decline. It is still able to strike a balance between economic recovery and social welfare. It is still the leader of high-tech sectors, such as IT, renewable energy, nanotechnology, and medical equipment. Although its government is "poor," its businesses are rich, and the dollar, the legal tender of the United States, enjoys predominance in the global money market. All this means that the United States will most likely emerge from the financial crisis ahead of other Western countries. In fact, since 2013, the US economy has entered a new round of revitalization, thanks to the general vitality of American society.

In 2000, Ren Zhengfei recommended an article to his colleagues: "Sleepless in Silicon Valley." From this article, you can really get a feeling for the core competitiveness of the United States. The article says:

> All those programmers, software developers, entrepreneurs, and project managers believe that sleep is akin to failure. Driven forward by their great expectations, and stimulated by giant cup after giant cup of coffee, they manage to stay awake in front of their fluorescent screens until four, five, or even six in the morning, far removed from the alluring comfort of their beds at home. This is the price they have to pay for their participation in the international market that spans different time zones. They have a new starting point each day, and they are fervently pushing the frontier of Internet technology forward.

> This is the uniform lifestyle of about 200,000 high-tech workers in Silicon Valley. Almost all of them stay up late. While most people follow traditional schedules of clocking in and clocking out at routine hours, in high-tech campuses the car parks are still crowded until 3 AM. Some high-tech workers may stay home at night, but they are typically connected to their office network on their personal computers.

> The high-tech sector, like sports, belongs to young people, but that also depends on how fast people age. Statistics show that most high-tech workers are single, male, and under the age of 35. Some try to squeeze as many products from their brains as possible to get more from their employers before their minds turn slow.

It's interesting to note that the mattress culture is not a unique creation of Huawei. It has precedents in Silicon Valley. For instance, the same article says: "At first, Netscape set aside a carpeted room where its employees used to sleep when they worked late at night. The company later called it off and encouraged people to go home and sleep in their own beds. But the employees kept asking the company to set up the rooms again." Similarly, at Huawei, has anyone heard complaints about the mattress culture? Today, mattresses of all different shapes and colors are still found under the desks in Huawei's offices.

In 2001, Huo Dawei went to the Huawei Research Institute in the United States on business. It was his first time visiting Silicon Valley, so he was eager to experience what this high-tech hub was all about. It was at a time in the United States when the IT bubble had burst and many start-ups had gone under—Silicon Valley had turned into a desolate valley. Despite this, when a local scientist took Huo to Cisco's parking lot at 11 PM, he saw many cars still parked there. The scientist told him: "There are cars here. Everyone must be working overtime." What impressed Huo was that when he explained Huawei's values to local staff, he was anticipating that they wouldn't understand; to his surprise, most of the local staff responded, "This is common sense in Silicon Valley. Businesses have to center around their customers, and success only comes through hard work. That's just how it is." Huo then realized that Huawei's culture was perfectly compatible with that of Silicon Valley: Huawei's culture wasn't unique to Huawei at all.

A senior executive at Huawei said:

> People often say that foreigners can't understand Huawei's culture, but I'm not quite sure why. Foreign companies are no stranger to customer centricity. They work on solutions to meet customer needs, and you have to be customer-centric when developing solutions or you won't get any contracts. This is pretty much a universal concept, and it's easier to understand if we put it in another way: Why are some people paid more? Because they work harder and contribute more to the company. More work for more pay, that's it. Recognition of hard work and contribution is the message we try to get across,

and I see no reason why anyone would have trouble with this concept.

Huawei now has 170,000 employees, among which 40,000 are non-Chinese. They are from different educational, national, and religious backgrounds, with different personalities and beliefs, but for the past 20 years, across all five inhabited continents of the world, they have been running together towards the same goal under the leadership of a handful of culturally distinct Chinese entrepreneurs.

What is the bond that holds them all together? A common set of core values and a common vision.

Customers: The Opiate of Huawei's People

On January 6, 2014, in a Chinese restaurant in Pudong, Shanghai, I interviewed Xia Zun, deputy general manager (DGM) of Huawei's Shanghai office. Half a year ago, he was the general manager (GM) of Huawei's Libya office, stationed there with two other Chinese colleagues throughout the war. After the war, he was transferred to the Shanghai office.

Me: What was the most unforgettable experience in your seven-year stint in Libya?

Xia: Once during a project negotiation I didn't go to bed for five days straight. I was 26 then, extremely energetic. On the day of negotiation, I didn't have time for breakfast, and the client allowed only one person from our side to go in the room. I went and negotiated from 10 AM to 4 PM, six hours straight without any food or drink, till the client suggested a break. I quickly went out for a Coke, a hamburger, and a cigarette. Then we continued until 6 PM, when the client eventually gave in. He said, "You guys are so spirited that I have nothing to worry about." The negotiator the client hired was a European, very skilled and coercive. If I wasn't able to make a deal then we would have suffered a huge loss. We signed a project for more than US$40 million that day, with an additional US$30-plus million at the end of the year, for a total of US$75 million. In that year we scored over US$100 million in that market.

Me: Were you rewarded after winning the US$40-million contract?

Xia: I got promoted to account director a few months after we signed the deal. In overseas markets back then, all positions were open, and those who ran the fastest could fill them.

Me: Where were you on the eve of the Libyan conflict?

Xia: On February 17, the turmoil began in the east and spread further on the evening of the 19th. I was attending MWC (Mobile World Congress) at the time in Barcelona. When I heard news of the conflict, I immediately flew back to Tripoli on the 20th. When transferring at the Amsterdam airport, there were only two Libyan locals and I, a foreigner, waiting for the flight. The flight attendant tried to confirm with me several times whether I really wanted to head back to Libya. That night in Libya, in response to the company's decision, the five of us on the management team held a meeting to arrange staff evacuation in four batches. And we agreed that the five of us would be the last to leave; whoever didn't agree could leave the room.

Me: Tell us about the process of evacuation.

Xia: After the discussion that evening, we sent the first batch to the airport on the night of February 23. The airport was in complete chaos. Afterwards, when I was in the office at 1 AM, I got a message saying that some seats were still available on the plane, and I sent a message to inform the second batch to get ready for evacuation. We were given 15 seats, but we had 40 people in the second batch. I led the efforts to get everyone from their dorms to the airport. When I got to the dorm, many of our team members were packed and ready. I called off the names of those who were to go first, and everybody was very cooperative, without a single complaint. The streets were dead by then and smelled of white terror.

Our local driver was very good. I am very grateful to him. The three of us—the driver, the DGM for delivery, and I—each drove a car. The driver insisted that he drive in the front. We wanted to give him US$3,000 before we left, but he refused. He said that he sent us off because he had grown fond of us, not because of the money. When we arrived at the airport, we found it was impossible to get in there—not because there were no aircraft available, but because the place was swamped with tens of thousands of people. We were relieved when our staff finally boarded the plane around 4 AM, but we dared not travel in the middle of the night, so we stayed at the airport till dawn.

The second and third batches had 58 people in total. Our company chartered an ISOS (International SOS) plane for us, but we still weren't able to get into the airport. In the end, a counselor from China's Embassy in Malta managed to get us in. The counselor's wife was from Beijing and helped us a lot during the evacuation. However, only 48 out of the 58 people got in. The company was willing to do whatever needed done—to fly out to Malta, each air ticket cost US$6000, non-refundable, and it was not guaranteed that we were able to board the plane. After a few rounds of buying and canceling, the company spent a total of US$1–2 million on tickets.

With the help of the Malta Embassy, the last batch left Libya by boat. It was midnight when we boarded a small boat that held about 500–600 people. I felt much more at ease once we were on the boat. But the wind was strong and the waves were too big, so the boat did not leave port until 4 PM the next day. We arrived in Malta at midnight. Two-thirds of the luggage space on the boat was filled with food and water to get us through the journey. Once we were in Malta, ISOS staff hired by our company came to take us to the hotel. The people at the Malta Embassy commented that Huawei arranged everything so well that they didn't have to worry. We kept order everywhere we went, always perfectly in line. Individual travellers at the dock and the airport followed in our footsteps.

Me: Why did you go back to Libya later on?

Xia: The company wanted to keep the flag flying in Libya. Our client called to say the network was down. While we withdrew to Tunisia, we received instructions from the company to return to Libya. We did a stress evaluation, and then flew back via Egypt. From our perspective, we only serve our customers, be they government troops or rebels. For us, customers are business groups. We also serve the common people—they are the ones who suffer most when the networks were paralyzed, because in times of war people need to report back home more than ever.

On March 16, the day before all flights were stopped, we took the last flight to Tripoli and called on our local staff to contact the customer. They were extremely moved, and signed a US$2-million maintenance contract with us soon afterwards. All of our local employees who could be found came back—at the time, Huawei was the only employer in Libya who could pay LYD500 cash salary.

Me: Tell us how you felt during the war.

Xia: I was a bit excited when we first got back to Libya, and would go to the rooftop of our building to watch. As time went on, we could only contact clients via telephone. Worse, our stored food grew mold when the weather got warmer. The local power plant ran out of petrol to generate electricity in the area, and we didn't have water either. Fortunately we saved a lot of water and gas back in January.

Me: What did you do every day during those eight months? Network maintenance?

Xia: Local staff were doing the maintenance, and so were our customers themselves. We were there to provide emotional support. Seeing clients once in those days was worth seeing them a hundred times in days of peace. However, as the war went on, the initial excitement was gradually replaced by mounting anxiety. I dared not go to bed till the air raid was finished for the day. Being trapped in the room like this every day was like serving prison time. There was no business either. This situation lasted more than two months, and I started to suffer from depression. After that experience, every time I hear the sound of firecrackers, I can't help but think it's gunfire.

Me: What else impressed you during the war?

Xia: We had a local staff member who helped us with a visa issue. We did not see him at all during the war—turns out he became commander in chief in the defense zone after the revolution. Another local staff member joined the revolution and was guarding the airport. He told us that he could help us take the airport VIP channel when needed. There were many unexpected things that happened. When a customer came to discuss business with us after the war, we were thrilled. All Huawei people are hooked on getting business. Business is like opium to us.

Me: You were stationed in Libya throughout the war. Did you ever think of bargaining for better terms with the company?

Xia: We always believed that the company would treat us fairly. After the war, I was promoted up two levels. However, I felt I hadn't done much for the company to deserve it. The promotion stressed me out; in fact, the stress was greater than the joy. Many people only see your promotion, but they don't see what you did for it. I took the initiative and suggested to my boss, Mr. Zou, that I stay in Libya until June to create more value for the company.

Me: Did you tell this story to your family?

Xia: Yes, after I came home. They told me I was courageous.

In my later days in Libya I realized that life was indeed fragile. I used to shop in the supermarket next to where I lived, and I knew a teenage kid there. After I came back from that evacuation experience, he wasn't there anymore. His brother pointed at him in a picture, indicating that he had died. This experience shook me to my core.

Me: After realizing life was fragile—and you mentioned that you did falter at some point—why did you still decide to stay?

Xia: It was a responsibility. I had been promoted, and I thought I should stay to earn that promotion, but as I stayed on I realized that my decision had nothing to do with the company any more. It was entirely a sense of personal responsibility that kept me in Libya.

The Sieve Theory: A Foundation of Dedicated Fighters

Employee Shareholding: Corporate Nuclear Fission

Commenting on the United States in the 1940s, Winston Churchill said, "The United States is like a giant boiler. Once the fire is lit under it, there's no limit to the power it can generate."

It's true. More than 300 years ago, the Mayflower brought a group of people with courage and dreams to the new continent. These fore-fathers of the United States brought along with them a sound social system of democracy, the spirit of contract, and the rule of law, which was based on Puritanism and valued individual gain and personal value. This system includes all necessary incentives for the country's growth and prosperity. For example, they have introduced a variety of employee reward schemes, including stock options, which have played a significant role in the rapid growth of high-tech companies in the United States. There is now general agreement that Silicon Valley has been driven by two engines: One is the stock option scheme and the other is its innovative culture; one provides material force and the other provides spiritual force.

The whole world is learning from the United States, and now China's high-tech, financial, and cultural industries are learning from their US counterparts.

Since its establishment, Huawei has placed a silver handcuff on its employees—the employee shareholding scheme. This scheme is referred to as "silver handcuffs" because it differs from stock option arrange-ments, or golden handcuffs. Although the two are comparable, Huawei's shareholding scheme wasn't adopted from the West. It was the company's only choice at the time.

In an article titled "The Spring River Flows East," Ren Zhengfei explained why and how Huawei developed its employee shareholding scheme:

> I designed the employee shareholding scheme soon after I founded Huawei. I had intended to knit all my colleagues together by a certain means of benefit sharing. At that time I had no idea about stock options. I did not know that this had been a popular form of incentive for employees in the West, and there are a lot of variations. The frustrations in my life made me feel that I had to share both responsibilities and benefits with my colleagues. I discussed this with my father who had learned economics in the 1930s. He was very supportive. But no one had expected that this shareholding scheme, which came into being by chance more than by design, would have played such a big role in making the company a success.

The scheme was Huawei's only option for competing with international giants and Chinese SOEs. It had no connections or resources as a private company, and it lacked funding. Everyone had to fight arm in arm with each other. Everyone had to own their own piece. Ren Zhengfei said:

> Please don't assume that I was a sage or a saint. If I had started a real-estate business, I would have taken it for granted that everything was mine because I got the land and loans through my own connections. Why should I share anything with anyone else? I would enjoy all of the benefits because I assumed all of the risk. But my company, Huawei, was established as a technology company. I needed more smart people, more people with lofty ideals, to work together and share the company's wealth and woes. Everyone had to take their own share of duty, particularly the older generation and the senior managers. Only when they diluted their equity could we encourage more people to join the battle.

Of course, you can't overlook the importance of the historical backdrop when Huawei was founded. In the 1980s, "capital" was a very sensitive word in China. I remember when studying psychology at the Northwest Normal University, a college politics lecturer had a vehement dispute with two other associate professors: "China should absolutely refrain from advocating this idea of competition. Competition is the essence of capitalism. We are a socialist country." Such was the dominant opinion at that time. Mainstream media were no exception, saying things like, "According to Marxism, a company owner who hires more than eight employees shall be considered capitalist…."

To establish a privately owned company against such a backdrop was a huge political risk. Therefore, in the Guangdong province—Shenzhen in particular—the owners of private companies older than 25 years and those which operated at a considerable scale usually didn't hold a large portion of shares. There are more than 2,000 companies with shareholding structures similar to Huawei's, which serves to indicate the trajectory of the system's evolution throughout history.

With keen political sensitivity, Ren was quite clear what it would mean if he was the majority shareholder at that time in history.

Ren Zhengfei is Huawei's largest shareholder, holding a 1.4 percent share. The remaining 98.6 percent is held by Huawei employees. At the end of 2014, among Huawei's 170,000 employees, 80,000 had a stake in the company (external shareholders do not exist). This is the most scattered share structure in the world and a rare phenomenon in business history. What is the source of Huawei's strength, its "nuclear fission"? From interviews with more than 100 Huawei executives, it seems most people believe it's the employee shareholding system—everyone is an owner, fighting the fight together. When asked what they thought about Ren Zhengfei, almost all of them mentioned his dedication or the fact that he's not at all selfish. Bai Zhidong, the head of Huawei's Blue Team,[3] remarked: "Ren is a man of his word. This to me is the best possible measurement of a person. How many leaders in the world are truly good for their word? He said he would share equity with his employees, and that's what he's done for over 20 years, splitting things up among 80,000 employees, and leaving only a tiny portion for himself."

No matter how a leader packages himself, true authority comes from showing results.

Some argue that Huawei's employee shareholding scheme has benefited many, but will also be the company's downfall in the future. Just as all flowers have it in their nature to decay, in the event that Huawei's business slows down one day, if its growth begins to stagnate, or if it begins to suffer losses, its employees would have fewer dividends or no dividends at all and the company would lose the cohesiveness and dedication that got it where it is today. This is not an alarmist take on the situation. So what kind of system can ensure that a company's management and staff do not falter or collapse when the company faces a substantial business crisis? Do public companies have the mechanism to achieve that? Fundamentally, the conclusion is more or less the same: Development is a hard truth. Only a healthy, growing company can create sustained cohesion between its teams and

---

[3] As part of its ongoing effort to identify potential weaknesses, Huawei employs a "Red Team and Blue Team" method of self-reflection. Similar to modern war games, the Blue Team plays the role of a competitor, always challenging the Red Team and exposing its faults to help the Red Team improve.

individuals; in turn, the premise of a company's healthy development is an incentivizing HR policy and a healthy organizational culture.

One Huawei executive gave an example of a healthy organizational culture. When Huawei was entangled in a lawsuit with Cisco in 2003, pessimism pervaded the whole company. However, Huawei ended up winning that lawsuit, which gave it the confidence it needed to face a lawsuit with Motorola later on.

Today, Huawei's innovation and intellectual property have gained recognition worldwide. These days, if anyone gets it in their head to sue Huawei, no matter who it is, much more courage, deliberation, and preparation are needed. Through the process of litigation, Huawei has gradually grown from being timid and uncertain to confident and, eventually, invincible. A collective acknowledgement has also formed, that is, the company's values haven't wavered over the past 20 years and, strategically speaking, it has made very few mistakes. "Follow the leader to victory" became a mantra of sorts, their "success-oriented rule of law," with the idea that "we won over and over again in the past, will keep winning today, and keep winning forevermore." This collective determination to succeed is an important force that sustains Huawei's culture.

Of course, this success-oriented rule of law is still built on a foundation of healthy HR policy.

Readers are advised to pay special attention here—Huawei believes in the philosophy that dedicated fighters are the cornerstones of the company. It values those who fight for the company rather than those who hold shares; the employee shareholding system merely serves as an important incentive (not the only one, to be sure) to motivate its "fighters." Shares are used to reward past contributions.

Ren Zhengfei said, "Financial investors should be reasonably rewarded, but the fighters involved in our own version of the landing at Normandy, or who dug our own Panama Canal—they are the ones who shall get more returns." Ren continued in this regard:

> Huawei is determined to ensure the interests of our fighters. For those who cannot fight anymore and wish to retire, we want to ensure their interests as well. If we stop taking care of the people

who fought in the past, who would fight for us in the future? However, we should reward those who fight in the present more than those who have already retired, otherwise who would fight for us now? It's true that I will retire one day, and it makes sense that I should support policy in favor of retirees, so that you work harder and I gain more—but nobody here is stupid enough to accept that.

Ren Zhengfei's insight into human nature is clearly deep. If Huawei was indeed built to last for centuries, then none of its systems should be set in stone. The internal mechanisms of an organization always tend towards disorder. Therefore, when leading a group of people, a company's mission should account for human nature and the capriciousness of time and space, while keeping a vigilant eye on disorder and adapting accordingly.

Staying Private Is the Only Way to Dominate the World

At a luncheon in a famous club in New York, Ren Zhengfei met with a dozen top American businessmen, including Hank Greenberg, the former chairman of AIG, and Vincent Mai, chairman and CEO of AEA Investors. Asked why Huawei had not gone public, Ren answered:

> Technology companies need motivated employees. If a company goes public, a bunch of its employees will become rich overnight—millionaires or even billionaires. And then they will most likely lose their motivation. This is not good for Huawei; this is not good for our employees, either. The company will stop growing at such a pace, and the people will lose their collective drive. If employees get rich too young, they will become lazy, which is a curse to their personal development.

Huawei advocates an organizational culture that fosters a sense of mission for those at high levels, a sense of crisis for those at the middle levels, and a sense of hunger for those at entry levels. The top executives are highly paid and enjoy more dividends. To some extent, wealth is no more than a symbol to them. These people are in the minority: They are not motivated by material gain, having already completed their formative years of accumulating material resources. What urges them on is a sense of mission and instinctual passion for their career. In short, the force that drives the elite at the top of the pyramid comes from much deeper within.

On the other hand, the mid-level managers who fought their way up from entry-level positions are often promoted because they have shown dedication throughout their years of working for the company.

But for the overwhelming majority of employees, more work for more pay is the basic and most realistic motive. This is a universal law of humanity that reflects human desire, a law Huawei understands and follows in its compensation design.

Huawei believes that mid-level managers should have a sense of crisis. If they fail to organize and drive their team to fulfill its goals or if they are selfish and complacent, they are quickly replaced or demoted. They will be promoted again if they regain their passion and meet all qualifications. "Entry-level employees should have a sense of hunger" for bonuses, company shares, promotion, and success. Such hunger breeds wolf-like aggressiveness. Without it, any sort of lofty appeal would be pointless.

Back in 1997, Huawei was already at par with its Western counterparts in terms of employee compensation. If this weren't the case, they would not have been able to attract and retain talent. This is also why, soon after it was established, Huawei adopted the employee shareholding scheme. The shares were not worth much at first, but they have now become the most valuable assets of its employees. Ren Zhengfei once said:

> Huawei has survived by virtue of hard work and technological innovation. Does technological innovation have no limit? Is Moore's Law always right? Can we keep our share of the market with one given piece of technology? I believe that technological innovation will slow down when the bandwidth and coverage of wired and wireless access reaches a certain level. At that time, only companies with extensive market penetration and excellent management that can provide high-quality services at low cost will be able to survive the cutthroat competition. So, Huawei

should try to reach this level before it dies off. Within the coming decade, we will try to learn from Western companies how to manage the business, how to improve its efficiency, and adopt a fitting human resources management system to motivate our people to fight for survival.

Ren has told the truth of the global IT industry. It is a cruel reality. Unlike other industries, IT is engaged in a sprint towards death that started very early on and will last far into the future. The law, however, is that the one who lasts the longest wins, no matter how fast they are running. Then how can a company avoid premature death? There is no choice but to fight. And how can a company turn its employees into motivated fighters? The answer is having a reasonable and superior human resources mechanism.

It's practically a natural law that people run faster if they gain riches in measured steps, but stumble if they get wealthy overnight. A lot of companies, both Chinese and foreign, are promising and robust before they go public, but they suffer from organizational shock quite soon after they get listed. And the reason for that primarily rests with the people who get rich too quickly: They stop pushing forward or they join their competitors for even more money. Some sell their share of the company and set up their own operations; worse still, they poach people from their former company and become the bitter enemy. Clearly, this system of managing human resources leaves something to be desired.

Ren Zhengfei once privately made the assertion, "If we do not go public, we might someday take over the world." This assertion has at least three layers of meaning.

The first refers to the team's collective fighting spirit. Abundance corrupts individuals and organizations. Too many "pies" will eat away at the very essence of an organization, which is the worst form of death for an organism. If the company remains private and its remuneration is on par with global industry standards, plus yearly dividends, it will not only attract and retain talented people, but also serve to maintain their will to fight. Up to now, Huawei has struck a successful balance.

The second meaning is that decision making should always be kept under control. Given the scattered share structure of the company, any private-equity investor could easily gain relative control over Huawei. And should the company become controlled by an investor seeking only short-term returns, Huawei would not be very far from collapse.

The third layer of meaning has to do with the goals of Huawei's people. Huawei has gotten where it is today, and has been able to overtake some of its Western counterparts, because it has a long-term vision. The company makes 10-year plans, while its competitors struggle to follow near-term fluctuations of the capital market. The capital market, so to speak, is a cold-blooded and impatient animal.

For example, Motorola had invested US$5 billion in its Iridium plan and when this plan failed, the capital market dumped Motorola's stocks and the company has been going downhill ever since. Similarly, Huawei invested nearly CNY6 billion in 3G products and for a long time there was no return on its investment (in some cases 3G products were sold as 2G ones). Nevertheless, Ren Zhengfei insisted on his vision, going against opposition in the company and prohibiting the development of PHS products. If Huawei had been a listed company, what pressure would shareholders have put on them to stray from their long-term vision? The conclusion is clear: Either Ren Zhengfei would have stepped down early on or Huawei would have passed its prime long ago.

A scholar offered the following insight (and Ren Zhengfei himself sometimes agrees):

> If there was an unlisted company among Huawei's Western competitors, Huawei would not be where it is today. In the immediate future, China simply can't produce business strategists that rival those in American, but great as they were, these American strategists had to succumb to the will of the capital market.

So when will Huawei go public? Ren Zhengfei repeated at the shareholders' meeting in April 2013 and at two press conferences in 2014 that in the following 5–8 years or even longer, Huawei would not consider going public nor consider engaging in any capital operations, including mergers or acquisitions.

The Sieve Culture: Simplicity Forces out Complexity

Two Chinese companies, Lenovo and Huawei, share an amazing similarity: The age gap between the founder and the top executives is over 20 years. This is a natural "father and son" or "mentor and protégé" age gap. Did this age structure play any special role in the growth of these two companies?

In its formative years of hunting and gathering, Huawei had assembled a group of ambitious young people. With them, there was hardly anything this half-starving company couldn't do. The harsh external environment, poor resources, and all kinds of internal management challenges couldn't destroy their will, serving instead to strengthen the cohesion of its employees. Undoubtedly, Ren Zhengfei played the role of a mentor, the soul of the organization—its flag bearer. An executive recalled: "It was like a madman leading a group of fools. With a wave of his hand, we charged towards the most unlikely of battlefields… He represented our interests, so we trusted him."

The president of Huawei's chip company recalled:

> Our boss was over 40 when he started the company. His followers were a bunch of simple, young college students in their early 20s, and he taught them a piece of straightforward knowledge—'treat customers well, and you'll get along just fine.' Most people believed in this, and found that money could be made in a legitimate way, so they followed him along this path.

Of course, for all of this to happen, they had to be "foolish," to be believers; once a believer, a positive mindset would naturally follow; and with the right set of values, the company was able to succeed while enabling every individual to succeed. On the other side of the spectrum, most of the early bunch that had their heads screwed on tight were skeptical, and skepticism led to divergent paths. After 27 years, Ren Zhengfei concluded:

> In hindsight, all the smartest people have actually left the company. The rest of us, the pedestrian among them, are still here. I'm like Forrest Gump: I believed that staying foolish could

lead to success, and I stayed. You're foolish as well, so you stayed. We succeeded in the past because we were foolish fighters. Today, foolishness will lead you to success all the same. The most defining characteristic of Huawei's culture is foolishness: A foolish person is customer-centric. We work together and fight together along the path we've chosen, each of us reflecting constantly to make ourselves better.

Someone asked Ren: "When do you think you started maturing?" Ren thought for a bit, and then answered, "Probably at the age of 57, when I realized the importance of interpersonal relationships."

So what's the most important thing in Ren's eyes? Customers and the value of customer centricity. When customer centricity was placed at the very core of its core values, Huawei's culture was defined: only those who are working for the benefit of customers and contributing to their success are entitled to monetary reward and promotion.

At the end of 2012, Huawei won a bid worth US$1 billion in another country, an achievement that the company had been working towards for many years. The bidding team was rewarded with a bonus of CNY7 million. When the contract was signed in the first half of the following year, Ren decided to reward the team an extra CNY10 million. The team leader, regional president at the time, said to Ren, "You've already rewarded them. This time, treating them to a nice meal is more than enough."

Ren got extremely angry at this comment, saying, "You don't care if your brothers are starving after you've fed yourself? I'll treat you to dinner. How about giving me your bonus, stocks, and salary? I'll buy you meals every day!" He called the same person five times in a row: "You really need to think about this seriously. Our brothers and sisters in the field offices have worked very hard, and they deserve their portion of the gains."

The regional president replied, "CNY10 million is far too much, we can't accept it. It's stressing me out—who gets that large of a reward for winning a project? I won't touch a penny for myself, but I still feel guilty accepting it for my team."

"Then tell me how much is appropriate."

Eventually, the sum was set: a second sum of CNY7 million.

"If we ask for a bowl of rice, he won't just give us a bucket of rice, but 10 buckets. If you are preparing for a big meal, he gives you 10 gold bars." Zou Zhilei, president of Huawei's Carrier Business Group, went on to say,

> Don't try to evaluate what our boss does in the short term. If you take a look at things from a few steps back, there are very few people who can do better than him. He doesn't care about your approach to a project; he's only concerned about results, and will support you with policy and resources, telling you that you earned it.

In essence, Huawei's culture is very simple, but it runs deep. Can you copy it? Of course you can, as long as you stick to the same principle of customer-centric value creation and as long as you evaluate performance and distribute value based on the dedication and contribution of your employees. But on the other hand, you'll also find it hard to copy, because in reality many organizations can't stick to such a one-dimensional culture, especially in any sort of sustainable, long-term way.

Organizational culture is like a coin: If the value on one side of the coin is different from the value on the other, everyone would be confused. Zou Zhilei sums up Huawei's view on talent management with the following phrase: "At Huawei, the principle behind everything we do is to keep things simple."

The hard workers and contributors will get rich—from day one to the present, this principle has been carried out thoroughly at Huawei. In the history of the company, there have been very few incidents of internal conflict, negative sentiment, or teams splitting up as a result of unfair distribution of rewards. At the same time, it's common practice for the hardest workers and contributors to get promoted. If you want a fast promotion, venture out to Africa, to tough places; if you want to lead, go for areas where the main battles are already underway. Huawei rarely talks about leadership training; what Ren advocates is a "horse-racing philosophy"—the commanders of armies are trained in the heat of battle. After all, who trained Mao Zedong? And who trained Deng Xiaoping?

The president of Huawei's chip company commented:

> Huawei doesn't engage in speculative opportunism—it has a grand dream. People are inspired because most of us are a bit idealistic inside. What the company believes is simple, so those who like to play tricks or politics don't survive long here. Those who inevitably stand out are the candid, hardworking contributors.

Rotating CEO Xu Zhijun said:

> Why is it easiest to select leaders during war time? Because war is cruel. First you cannot die; if you didn't die, it means the team you led won the battle. Whereas if your team lost the battle, you yourself wouldn't have survived. Heroes are produced by the times, and leaders who emerge during war time are those who are truly identified or selected based on their performance. [...] Selecting who to promote reveals a sort of orientation within an organization. In Huawei, all of our policies over the years have made people feel that, as long as you do your job well, you will have bright prospects, a strong platform for development. Therefore, focus on doing your job well, instead of building relationships and brownnosing. Save your brownnosing for the customers.

Historically, Huawei used to link leader evaluation to performance as well as what they call "key behaviors." The latter usually led to subjective observation, that is, if they felt that you were competent (without hard KPIs, or key performance indicators), you would be given a chance to prove yourself. The result was that, if your evaluators said you did a great job, you did a great job; if not, you didn't. This led to randomness and unfairness in leadership appointment. In 2005, the company's senior management went through many rounds of heated discussion to finally reach consensus. Moving forward, KPIs would be a watershed for leadership appointment: Only those among the top 25 percent of KPI achievers would qualify for key-behavior observation. Thus, from the system level, an end was put to what Xu Zhijun called "the Ho Chi Minh trail" phenomenon—finagling promotions through sycophancy, subterfuge, or forming interest groups.

A Huawei technical expert, who asked to stay anonymous, commented:

> Our boss always said that all the outstanding guys left the company and only the fools stayed—this is not true. Without excellent people, how could Huawei be where it is today? Why is it that those who stayed behind have a simple mind? Smart people do not need complexity. They gave their trust, they made a choice, and they stayed around to follow. Over the past 20 years the company was like a sieve. This sieve filtered out the people who were complex, but not necessarily exceptional; whereas people who were simple, but not necessarily unexceptional, were left here. The simple people in the sieve looked odd against the complicated backdrop of society. Huawei's culture, in essence, is a culture where simplicity forces complexity out. It's like a campus culture, where Ren is our mentor, or the schoolmaster and mentor in one.

Perpetual-motion Machines Injected with Rooster Blood

Just 12 minutes after takeoff, the plane started to rock violently and nearly entered a dive straight towards the ground. The passengers could see mountain peaks flying past the windows. Ren Zhengfei was scared, because his wife and daughter were right beside him. At this time, a stewardess announced that the aircraft had a problem and they were returning to the airport.

Several minutes later, the airliner landed at Beijing Capital International Airport. On the apron, a dozen police cars and fire engines were waiting for them. Their sirens were blaring, which further intensified the atmosphere. "Thank God! We are alive!" Ren Zhengfei said to himself. He was still extremely nervous; his face was pale. Yet his daughter, to his amazement, was rather thrilled. She seemed to enjoy the spine-tingling drama.

About 10 days later, Ren met with another near escape. This time, it was a flight from Cairo to Doha. All of a sudden, the aircraft ran into severe turbulence, bumping violently up and down. It was almost as if the pilot had passed out at the wheel. In the end, the aircraft landed safely at the Cairo Airport. Sitting on a bench in the airport terminal after this shocking event, Ren Zhengfei asked another passenger whether she had been afraid. "No," the passenger said, "I have seen patients dying in hospitals. Life is fragile. We must live in the moment and live every day to its fullest." Ren had thought of canceling the journey, but he was greatly inspired by this answer. Two hours later he was on another flight to Doha.

In addition to traveling the world over, Huawei's executives have kept their mobile phones on 24 hours a day for over years. Wherever they are, in China or other parts of the world, they must be ready to answer the phone. Huawei has a presence in more than 100 countries and now has more than 100,000 employees. Things happen every day. Ren said, "I don't want to hear good news. I don't care much about it. But whenever something bad happens, I must know it, especially things that concern the safety of our people. They are our fighters."

According to his close colleagues, Ren Zhengfei takes approximately 100 flights each year and other senior executives at Huawei, such as Sun Yafang, Hu Houkun, Xu Zhijun, Guo Ping, Xu Wenwei, and Chen Lifang take more than 150 flights each year. At the same time, more than half of Huawei's board and EMT members have suffered from stress-related diseases. Anxiety is the most common illness, then hypertension, diabetes, and depression. Guo Ping has it the worst: In the past, he has suffered from a condition with an indiscernible cause. When he gets sick, his blood platelets drop to an extremely low level, but after he gets some rest, everything is fine again.

What is it that keeps this group of people going? Ren's wife described her husband in this way: "For the past 20 years, it's as if he was injected with rooster blood every day." The rooster blood therapy that she jokingly refers to here was a popular practice in China in the 1970s. People would get injected with fresh rooster blood as a way of enriching their own blood, to rejuvenate them, return them to a state of youthful vitality. It comes as no surprise that this was later proved to be a scam—a nostrum. However, the phrase "injected with rooster blood" became widely used to describe extremely energetic people who think actively and work tirelessly.

In Huawei, there is a group of people, especially senior executives and researchers, including some foreign scientists, who all show signs of having been injected with the proverbial rooster blood.

What keeps them going? Their sense of mission.

When it comes to a sense of mission, after observing Huawei for over a decade, I found that there are three categories of people. The first type is a very small number of people who were born with a sense of mission. When they set a goal in life, they begin to think and act like a perpetual-motion machine. No matter the situation, despite any pressures or difficulties, they can quickly mobilize every cell in their body to march forward and face what's coming. There are indeed a very few politicians, scientists, artists, and entrepreneurs that were born with this innate drive. They are simply made of different stuff.

Steve Jobs, Ren Zhengfei, and Elon Musk clearly belong to this category.

The second type is what I call "mission-inert" people. They are driven forward by responsibility for long stretches of time that ends up forming a strong sense of mission. Their arduous diligence doesn't depend on some specific plan, but is instead an internal source of momentum that is derived from responsibility.

In 1990, Guo Ping was commissioned to develop 256-line switches. During project kickoff, Ren Zhengfei asked him, "How much research funding does your college research supervisor get in a year?" Guo Ping replied, "About CNY800,000–900,000."

Ren said, "Look, I'll give you CNY10 million—you make it happen."

Guo Ping felt what he described as "a sudden surge, a noble sense of mission and responsibility." He was determined to succeed in this task. He worked till midnight every day, often staying up in the lab all night long. When he finally finished the product, it was a 512-line switch. Up to this day, nobody knows how much money was spent on product development. Whether or not he spent the entire CNY10 million is anyone's guess.

Sense of responsibility held for an extended amount of time will transform into a sense of mission. Some people work day and night without feeling any happiness. Pain and pressure becomes a daily theme in their lives, the so-called "unbearable heaviness of life." However, if they are in a situation like this for too long and let pressure become a part of their daily rhythm, once they step away from an overburdened work environment, they'll inevitably feel "the unbearable lightness of life" and begin to crave the pressure and commitment all over again.

The third category of people is the "mission–passive" group. In an organization, most employee–employer relationships are contractual, where the employer dispenses responsibility and the employee shoulders it. An outstanding organization can always engage the maximum amount of employees in a collective sense of mission by widely promoting the company's values and strategy, effectively incentivizing its people, and building a more aggressive sense of culture. However, objectively speaking, to expect all employees to have such a sense of mission is neither realistic nor rational. Therefore, as mentioned above, Huawei advocates an organizational culture that fosters a sense of mission for those at high levels, a sense of crisis for those at middle levels, and a sense of hunger for those at entry levels. When trying to induce a sense of mission among entry-level employees, the company puts more stress on employee engagement.

Gallup's report on the "State of the Global Workplace: Employee Engagement Insights for Business Leaders Worldwide" highlights findings from Gallup's ongoing study of workplaces in more than 140 countries from 2009 through 2012. Based on responses to the survey, Gallup groups employees into one of the three categories: engaged (enthusiastically committed to their work and their organization), not engaged (put in time but lack energy and passion), and actively disengaged (unhappy about their work, actively showing their discontent by undermining the engaged ones).

The report shows that only "6% of Chinese employees are engaged in their jobs—one of the lowest figures seen worldwide."

This result is undoubtedly frustrating. A Google survey shows 99 percent employee engagement, while the average engagement rate in the United States is less than 20 percent.

Professor Larry R. Donnithorne, former director of the West Point Military School's Leadership Center, asked me about Huawei's employee engagement rate. I told him it was about 85 percent. However, when a Huawei senior executive learned about this response, his comment was, "That's way too conservative!"

Success, or sense of mission, is the personal perception or sense of value that each person possesses and depends upon. Without this sense of value, people feel empty. Emptiness is a dark force that can destroy a person. Throughout human history, prominent and successful people have substituted social commitment and lofty values to fight off the perception of emptiness. They are destined to lead a difficult life, filled with solitude and pain.

Ren Zhengfei and other leaders of Huawei carry the same cross on their backs. They are obliged to drive the Huawei express train to every corner of the earth, to keep it in perpetual motion, and to fulfill a sense of national pride. The greater the underlying desire, the stronger their sense of mission. If it is going to live long and thrive, Huawei needs to ensure that its people are properly motivated. The company should therefore develop more extensive and effective incentives to resist or offset (because it is impossible to eliminate) organizational fatigue. This is a big challenge for its leaders and managers. Although a perpetual-motion machine is only hypothetical, Huawei's leaders must keep themselves in perpetual motion. And while they themselves try to resist "leadership fatigue," they must lead the organization out of fatigue as well.

In this sense, not everyone can be a good leader, as it's one of the most painful jobs out there. It's not a mere issue of competence, personal integrity, or charisma. It's a question of how to overcome fatigue and boredom, which makes being a leader the most unnatural job in the world, the one that's most contrary to human nature.

Openness: A Matter of Life and Death

Chapter 3

An Era of Creative Destruction

"He Never Saw Huawei as a Family Business"

One day in 1993, Ren Zhengfei was strolling around Zhong Guan Cun, a technology hub in Beijing. Someone asked him, "What do you think of Founder?"

He answered, "Founder has excellent technology, but it lacks good management." At that time, Founder's electronic typesetting system was making short work of traditional Chinese-character typesetting technology.

"What do you think of Legend (later renamed Lenovo)?" he was asked again.

"Legend has excellent management but lacks advanced technology," Ren answered.

The inquirer went on, "What about Huawei?" To which Ren Zhengfei replied, "We have neither advanced technology nor good management."

Ren Zhengfei was telling the truth. This was the reality of Chinese IT companies at that time. Since then, however, they have made great strides to catch up with industry leaders. The year 1993 was an iconic year, not because of any major global event, but because 1993 marked an important milestone for the IT industry. This was the year when the Clinton administration announced its plan to build the National Information Infrastructure (NII), a network of information highways. NII marked the beginning of a new era: an era of creativity with a dash of destruction. After all, the cost of innovation and creation is the destruction of old things.

By 1993, Huawei was nearly six years old. With less than 400 employees, its sales in the first half of the year had just surpassed CNY100 million. That year, Huawei developed and launched the JK1000 analog switching system, but its market performance was discouraging. In the same year, Huawei's C&C08 2,000-line switch was under its first trial run in the Posts and Telecommunications Bureau of Yiwu county in Zhejiang province. The company had also begun R&D on a new C&C08 switching system with a capacity of 10,000 lines.

The year 1993 was the real starting point for Huawei, because it had begun to open itself up to the outside world. Over the past six years, the company's goal had been to overtake Stone, one of the leading Chinese IT companies in the 1990s. After 1993, however, Huawei began to vie for a place among the top three global telecom-equipment providers. Lenovo at that time was also weaving its dream of challenging IBM.

At the end of 1997, Ren Zhengfei led a delegation of Huawei's senior executives to the United States. They traveled across the country to visit a number of American companies, including IBM, Bell Labs, and HP. The history of the IT industry in the United States, especially the frequent rise and fall of companies, came as a great surprise to them. Ren Zhengfei said:

> One large corporation after another got caught up in trouble and died off; small firms burst forth into towering trees, which then got struck down by lightning. It's an endless cycle of birth and death, almost as if the 500 years of China's warring states period transpired within a single day in the US.

Nevertheless, Huawei's executive team clearly felt the enormous power of the open-minded culture and system of innovation in the United States. This was a country of heroes, each of whom may lead the way for decades, or for just a few short years, may be those who inspire the entrepreneurship and innovative power that underpin the strength of the nation.

On Christmas Eve, households all across the United States twinkled in the dark like stars. But Ren Zhengfei and his colleagues had shut themselves up in a small hotel in Silicon Valley. They had a meeting behind closed doors for three days, churning out a document of more than 100 pages. IBM's management transformation was a source of great inspiration for Huawei: A small company lacks competitiveness while large companies collapse if they are not effectively managed. Extremely hard work was endemic to the US high-tech industry, especially among successful entrepreneurs and senior executives. Millions of dedicated, hardworking people were the driving force behind advancements in technology and management, leading to the creation of many great companies in the United States. This resonated strongly with Huawei's people: Hadn't Huawei itself grown up on the shoulders of unwavering diligence?

Huawei's management was also shocked by the dramatic rise and fall of Wang Laboratories, established by Dr. An Wang, a Chinese-born American. In 1971, Wang Laboratories launched the world's most advanced word processor, the Wang 1200. By 1978, Wang Laboratories became the largest information product supplier in the world, yielding US$2 billion in personal wealth for An Wang. In 1985, An Wang ranked eighth on the *Forbes* list of the 400 richest Americans. But in 1992, he filed for bankruptcy protection and the company's share price plummeted from an all-time high of US$43 to 75¢.

What went wrong? The company wasn't open enough. Wang Laboratories had enviable R&D capabilities, but the problem is that, in an age of shared information, no single company or individual can cope with rapid technology shifts on its own. In the case of Wang Laboratories, a closed model for technological development partly explained the company's collapse. The fundamental cause, though, was the closed nature of its culture. In its heyday, Wang Laboratories had a great number of tech geniuses, but the company had been managed exclusively by the Wang family. Dr. Wang handed power over to his son and management soon became enmeshed in family politics. Eventually, the company suffered from a mass exodus of its tech leaders.

An interesting episode occurred in 1991, when a young man named Chambers left Wang Laboratories and later joined Cisco. When he passed the torch to Chambers in 1994, Don Valentine, former CEO of Cisco, told his employees, "Chambers will make Cisco invincible." Indeed, Chambers fulfilled this prediction. However, around the year 2000, Cisco met a new rival—Huawei. And Chambers met a challenger from the East—Ren Zhengfei. Of course, this is a story for later.

In 2006, at a meeting in New York, I asked an influential American investor: "What's your opinion on Wang Laboratories by An Wang?" The American was confused: "Who's An Wang?" After I explained, he said: "I'm afraid not many people remember this company. The USA is a forgetful place." I thought dismally that fame is nothing but a mercurial thing. It seemed as if even the last trace of An Wang, the once almighty computer pioneer and IT hero, had been erased from people's memories. Likewise, who would remember Zhang Shuxin, the "woman of the decade" in the 1990s, the first person to "eat crab" in the Chinese Internet sector?

I asked him another question: "If An Wang's successor was Chambers or someone else, would it have made a difference?" He answered, "It's hard to say. Maybe it would, because Chambers is a great businessman."

I first met Ren Zhengfei in 1999. The two of us were having tea with a middle-aged scholar at the café in Shenzhen's Wuzhou Guest House. The scholar spoke highly of the family-owned business management model of Chinese businessmen in Hong Kong and South-East Asia. He regarded it as the primary path to success for businessmen in Eastern countries. Ren Zhengfei became very impatient:

> I don't care what Li Ka-Shing or Cheng Yu-Tung[1] do. Whoever takes over at Huawei will have fought their way into the position. Whoever has the ability to win people's hearts, knows how to cooperate and compromise, and is open-minded and broad-hearted, will be our next leader. Furthermore, we work in groups, because two heads are always better than one.

---

[1] Li Ka-Shing is the richest man in Asia as of November 2015, and Cheng Yu-Tung is the billionaire owner of Chow Tai Fook Enterprises.

Essentially, as early as 16 years ago—maybe even earlier—Ren Zhengfei was determined not to let his children be his successors. For over a decade, the media had cooked up false news reports on family succession at Huawei, which the company's executives always brushed aside with a chuckle: "The media doesn't know our boss. He has world-class ambitions. If it means success, he's more than willing to divide up the lion's share of company stocks with everyone. From the very beginning, he never treated Huawei as a family business."

Since 2013, Ren Zhengfei has explicitly announced in various share-holders' meeting and media briefings, "My family members will never succeed me." In fact, other world-class companies in China, such as Lenovo, founded by Liu Chuanzhi, and Haier, founded by Zhang Ruimin, have all adopted the same principle for corporate governance, that is, appoint-ment by virtue and ability, not by blood ties. Therefore, the so-called "Eastern path to success" doesn't seem to hold ground in mainland China's business world.

The rise and fall of Wang Laboratories taught Huawei's management team a profound lesson: In this age of rapid technological and social change, a company can't stay the course if it keeps to itself in a closed space; companies have to advance along a path of openness in order to survive and grow and to catch up with their global peers. Anything else is a dead-end road. In a 1999 message to his new recruits, Ren Zhengfei stated with utmost certainty, "Huawei must keep learning from the outside world if we want to survive. We must open our door if we want to catch up with the world. We have developed each and every major product through open partnerships."

In 2012, the company further articulated the lesson they had learned over 10 years ago in the form of a board resolution:

> Huawei must hold true to its open policy in the long term. We must not waver under any circumstance. If we do not open our doors to absorb energy from the outside world, we won't get any stronger. At the same time, we must examine ourselves critically throughout this process; otherwise, we won't be open for long.

In short, openness is the basis of Huawei's survival, the source of its growth.

Learning from American Companies: Where's the Brake?

One funny story about Ren Zhengfei is widely known among Huawei's older employees. Near the end of 1997, Ren Zhengfei sold his second-hand Peugeot and bought a BMW. Whenever he was free, he would drive his BMW along Shennan Avenue, a main artery in the city of Shenzhen. He would open the sunroof and the windows, enjoying the cityscape and listening to English lessons along the way. One day, he drove past an old car that was moving rather slowly. He turned his head to discover it was Louis Gerstner, Jr., the CEO of IBM. Without saying hello to Gerstner, Ren Zhengfei asked, "Have you ever driven a BMW?" Gerstner did not answer. Some time later, when he drove back and past Gerstner again, he asked, "Have you driven a BMW?" He still got no reply. When Ren asked the same question a third time, Gerstner got a bit upset, asking, "What do you mean?" Ren explained: "Do you know where the brakes are on a BMW?"

Although a joke, this story has an interesting message: In 1997, Huawei was growing very fast. It had learned to step on the gas, so to speak, but it hadn't learned how to slow the car down. In other words, it knew how to grow but didn't know how to manage its growth properly.

This may not be completely true, though. Back in those days, learning to accelerate was far more pressing than adopting a system of control. Huawei still needed to grow up fast and—pushed forward by an indomitable spirit—rapidly expand its presence in the global market; it was not yet time for Huawei to step on the brakes.

This 1997 tour around the United States proved to be a pivotal event for the company. Since then, Huawei has introduced the process-based management of IBM and the experience of US companies has played a great part in its strategic design. An organization that learns, grows. Ren Zhengfei said, "The process of borrowing and adapting ideas is a good thing. Western management philosophies and technology have proven successful—why would we resist them? Huawei should first apply more rigid methodology, adapt it, and then finally institutionalize it. This is a process we must go through."

Ren Zhengfei has repeatedly noted that

> we must study advanced things in the US if we ever intend to surpass our American counterparts. Throughout this process, we must keep a select few politicians and the great American people separate in our minds. We should not give up the opportunity to learn from America out of distaste for a few individuals within its government.

Indeed, Huawei has tried to learn from the United States. Ren Zhengfei and other senior executives have made frequent visits to this end. No doubt, Huawei has been a faithful student. So then, why has the teacher turned on its student? What's the matter with the United States? And what's wrong with Huawei?

Some Western media outlets and politicians have consistently viewed China and Chinese companies through a lens of bias, accusing them of being conservative, closed, or mysterious. What they don't know, however, is that the primary reason behind China's massive progress over the past three decades is the entire country's spirit of openness as well as the open-mindedness of the business community at large. This is fairly apparent to anyone that has a decent knowledge of China's history of reform.

Merchants in China have historically been belittled. Business trade was an extremely undignified, even illegal profession. Therefore, China does not have a commercial system or culture like the West. The so-called "Chinese way" of business management didn't exist for the longest time, and is built on a shaky foundation. China has borrowed some business concepts from the West and has integrated them to a certain extent, but even the most successful companies, including Huawei, Lenovo, and Haier, have not yet built a system of their own on top of all the tenets they've imported. This is a tough truth, one of which insightful people in China's business circles are keenly aware.

While it's humorous to think that Ren Zhengfei has no idea where the brake is on a BMW, this joke carries profound implications. Huawei was open, but it was still groping for stones in the river, and that was risky: It stood to lose everything should it spin out of control. During this period, Huawei was facing a "Death Valley" that many Chinese companies have faced at one time or another after they had finished up their formative years of hunting and gathering and were floating through a period of system-less ambiguity on the way to becoming a modern corporation.

Would Huawei be able to navigate through this valley? Beyond that was the question of openness: If Huawei were to open its doors, to whom should it open them? Whom should it learn from, and what should it learn? By 1997, Huawei had developed its own management system, with some concepts and models based on its own experience, its acquired knowledge, and the wisdom of its leaders. Yet this system was simple and locally oriented, and the concepts and models were merely slogans. The system proved unfit for internal management and external expansion. At a large meeting with his management team, Ren Zhengfei asked if anyone had ever managed a company of CNY10 billion in sales revenue. No one answered, and he went on to conclude that the only way ahead for Huawei was to learn from those who had, in particular its US counterparts.

In late 1997, Arleta Chen, a senior executive at IBM, gave Ren Zhengfei a book about integrated product development (IPD) transformation, which marked the beginning of Huawei's education on enterprise management from an American company. Since that time, IBM has helped Huawei build up a complete set of modern business management systems, processes, and even culture. In doing so, IBM has earned a considerable amount in service revenues and has managed to promote its own corporate culture and organizational philosophy in the process. Having introduced IBM's management consulting into all aspects of its business, Huawei has undergone dramatic organizational transformation and has learned how to control the power and braking system of the BMW—its full set of functions. Huawei has eliminated chaos and disorder from the organization and has laid a solid ground for an organizational and corporate culture that is more East than the West, even more West than the West as it enters the international market.

This is exactly the power of openness. Without strong courage and the determination to open up, Huawei could not have become what it is today. In contrast, a lot of Chinese companies have also commissioned advisory services from leading American consulting firms, such as McKinsey and Accenture—and at a high cost. But few have reaped the benefits. Some of them, as the joke goes, managed to die off even faster after taking the pills prescribed by their foreign doctors.

Why? Naturally, there is the problem of cultural differences, but the key problem is that, despite hearty claims to the contrary, their decision makers are not really committed to openness. They lose heart midway when difficulties arise. Beyond courage and spirit, possessing the necessary wisdom to manage the reform and opening-up process is critical.

Hunting Huawei: What's the Matter with the United States?

On February 25, 2011, Huawei's newly appointed deputy chairman, Hu Houkun, published an open letter addressed to the US government to clarify false allegations in some Western media, including "close connections with the Chinese military," "financial support from the Chinese government," "disputes over intellectual property rights," and "threats to the national security of the United States." The open letter also invited the US authorities to conduct a formal investigation into any concerns they may have about Huawei.

The allegation of military ties is nothing new. These claims are based entirely on the fact that Huawei's founder and CEO, Ren Zhengfei, once served in the PLA's Engineering Corps in a technical position equivalent to that of a deputy regimental chief. This "coming out of the military" is not rare in the United States, where a significant number of chairmen, vice-chairmen, and CEOs of Fortune 500 companies graduated from West Point. West Point is not only a military academy but also an incubator for business leaders: It has produced as many industry captains as all business schools in the United States combined. After World War II, the three major military academies—West Point, the United States Naval Academy, and the United States Air Force Academy—have turned out over 1,500 CEOs, 2,000 presidents, and 5,000 vice presidents at Fortune 500 companies, as well as thousands of leaders at small and medium-sized enterprises. Does that mean their companies are connected with the US military?

The allegation of the Chinese government's financial support originated from a speech by Cisco's John Chambers at a summit in Europe, where he attributed his company's dismal market performance to the US$30 billion financial support that Huawei supposedly receives each year from the government. Huawei sent out a letter, kindly worded, to Chambers and explained, "For the past decade, Huawei has engaged KPMG, a US-based global accounting firm, to audit its financial statements. We would also like to know where this alleged US$30 billion came from, and where it has been spent." Huawei issued similar statements through media outlets in Europe.

In early 2013, the US House Permanent Select Committee on Intelligence issued a report on Huawei. Except for the question of why a private company has a "Communist Party Committee," most of its other accusations were criticized by major US and UK media as "far-fetched and feeble" and "full of American imagination." The former Party secretary at Huawei, Chen Zhufang recalled that, at the time when the Huawei Party committee was formed, she went to the US-invested Motorola (China) to learn from their Party Committee. According to a media report, the German GM of the Sino-Germany joint venture Shanghai Volkswagen strongly supported establishing a Party Committee in the joint venture because he believed it would enhance employee cohesion and improve labor relations. As a Chinese company, why can't Huawei form a Party Committee? In fact, Huawei's Party Committee (which is referred to as the committee of ethics and compliance overseas) plays a rather important role in governing morality and social responsibility among business leaders in the company.

Since 1993, the world has undergone a number of disruptive changes without historic precedent. And the United States, by virtue of its open and innovative culture, has brought the human race into a state of an irrecoverable past and an unpredictable future. In the past 20 years, through creative disruption, the United States has produced a number of legendary companies such as Google, Yahoo, Microsoft, Cisco, Apple, and Facebook, and its strength is derived from hundreds of super corporations that are richer than some countries.

US society believes that the country's core competitiveness lies in its businesses. Meanwhile, the people of the United States, who hold a strong faith in empirical and pragmatic philosophy, also believe in the law of natural selection or survival of the fittest. Therefore, over the past two decades, innumerable US businesses, including some giants, have fallen, collapsed, or disappeared amid the tempest of technological change.

Like Cisco and Google, Huawei is an emerging company that poked its head above the horizon a mere 20 years ago. It has risen by learning from US companies and, in this process, it has witnessed one great behemoth after another falling down and out.

Starting in 2003, when Cisco first filed a lawsuit against Huawei, a number of US companies have also taken their own shots at the company while the US government and media have accused the company of posing a national security threat and engaging in unfair competition. Later, their attempts at suppressing Huawei intensified and the aim was clear: They wanted to banish Huawei from US territory. Behind all the smoke and mirrors, they were engaging in trade protectionism to defend the market position of US companies.

Since 2007, Huawei has fallen victim to a nonstop onslaught of false allegations from Western media. The most typical story starts with Huawei's dubious background and its sales of equipment to Iraq and Iran, which threaten the national security of the United States. One of the most absurd claims was from the *Wall Street Journal*, asserting that Huawei had assisted the Iranian government in tracking dissidents. Huawei had issued an official statement of denial, condemning the report as false, deliberately misleading, and patently untrue.

In 2007, 3Com Corporation announced a definitive agreement to be acquired by Bain Capital, a leading global private investment firm. As part of the transaction, Huawei would acquire a minority interest of 16.45 percent in the company and become a commercial and strategic partner of 3Com. But this deal was resisted by several congressmen and denied by a very high-ranking official in the federal government.

In 2010, Huawei's attempted acquisition of a US$2 million US firm was also rejected by the US government.

Once again, in 2010, Huawei offered to buy Motorola's wireless network business at a premium over the leading contender, but this bid was denied by some members of the US Congress. In fact, Huawei had been the original equipment manufacturer (OEM) of Motorola's wireless products, where Huawei's proprietary technologies were being used.

In the same year, five US senators tried to bar Huawei's bid to supply equipment to US operator Sprint. Throughout the process, the mainstream US media continued fanning the flames.

What's wrong with Huawei? As a retired senior US politician admitted, Huawei has successfully caused stress and fear among its competitors as it continued to expand its presence in the global market.

One morning in August 2012, I was having breakfast with Ren Zhengfei at the Ritz-Carlton in Beijing when a middle-aged man in a neat suit approached us. He presented his business card, saying that he was an executive with Citigroup China, that he was discussing a financial deal with Huawei, and he needed Ren's support. Ren replied:

> Sure, but I suggest you go to Guo Ping directly. He is responsible for all financial affairs. But you guys in the US are sometimes too closed, politicizing almost everything that comes your way. The UK, on the other hand, is a lot more open, far more open than the United States.

A State of Void Inflicts a Spirit of Openness

Openness Is a Way of Thinking

In 1954, Albert Einstein pessimistically predicted: "The unleashed power of the atom has changed everything save our modes of thinking, and thus we drift toward unparalleled catastrophe." In retrospect, his prediction has been only partially correct. Over the past 60 years, the human world has gone through quite a few major crises, including two financial crises triggered by the United States. Yet we have survived. Why? Because we have changed the way we think. Einstein had not predicted that the way we think might change as a result of external circumstances and that we are always able to detect opportunities, even in a crisis-stricken world. This gives us the capacity to keep moving forward.

Huawei has grown by leaps and bounds in the past two decades because it embraces change with an open mind. On the other hand, some large and established companies have been endowed with the nobility, pride, and confidence of a successful past. They cling to an almost innate paradigm, one that effortlessly speaks of tradition. They don't care about change. As a result, they have been overtaken by new start-ups, such as Google, Apple, and Huawei, who do not have a long history or tradition. A company may be underprivileged if it has no roots in tradition, but this lack of tradition may turn out to be an advantage in an age when external changes outpace internal efforts to reform. Free from any restraint, a company without tradition can easily break away from old rules and participate in the formulation of new ones. With an open mind, Huawei has developed a unique culture that combines both Western and Eastern characteristics. This is one of the secrets behind its rapid growth.

Ren Zhengfei is also an open business thinker who embraces the West and the East, the modern and the traditional. He is a born student, with an extraordinary capacity to learn new things. While watching a TV series, *The Qin Empire*, Ren Zhengfei was deeply impressed by Shang Yang, one of the most tragic reformers in the history of China. He was awestruck but also felt regretful. He believed that Shang Yang had chosen the right path, but his approach to reform had been too radical. Radical reforms often come at too great a price.

Among foreign politicians, Ren appreciates Yitzhak Rabin the most, calling himself a student of the former prime minister of Israel. On the other hand, Ren regards Ariel Sharon as a shortsighted politician with hawkish diplomacy. Ren Zhengfei believes that the policy of giving "land for peace" proves the wisdom and vision of Yitzhak Rabin, and the thinking behind it has become a principle for corporate governance at Huawei.

In 2009, Ren Zhengfei told the story of the Last Supper at the Pentagon. He said,

> Now the financial crisis is roiling the market, and the future is unknown. But I don't want our boat to sink. Given that we're all in one boat, we have to keep our direction clear, we have to stay the course, and row as hard as we can.

The Last Supper at the Pentagon refers to a famous dinner offered by William Perry, then Deputy Secretary of Defense, to the bosses of arms suppliers in 1993. During the dinner, William Perry warned his guests that the budget was going to shrink and consolidation was essential for their survival. This dinner led to a rash of industry-wide mergers in the United States beginning in the mid-1990s. Of the 50 major weapon vendors in the 1980s, only five survived in 2002, including Boeing and Lockheed Martin, which are multi-sector conglomerates.

Ren Zhengfei has always been an excellent learner. Thirty years ago, he was recognized as a model learner of Mao Zedong's works, which have undoubtedly served as a source of inspiration. In the media, some critics say that Huawei's management philosophy bears the mark of Maoism, and they do have a point. However, if they think that Ren Zhengfei is using Maoist thought to run Huawei, then they are gravely mistaken.

Among Chinese political leaders, Ren Zhengfei has the greatest respect for Deng Xiaoping. He has asserted on multiple occasions that Deng Xiaoping is the greatest reformer in the history of China and that Deng's ideological legacy consists of two key words: "reform" and "open." These two words are precisely the recipe for Huawei's success.

Since it was established, Huawei has defined itself as an open organization. Ren Zhengfei strongly felt that Huawei would not survive if it were to act as a fortified village. He said,

> Openness is the basis for the company's survival. If we do not open up, in the end we will only perish. To be open, we must grow our core capabilities and strengthen our partnerships. Huawei must not stray from this course, because openness is our only option.

When it comes to cross-cultural management, Huawei executives' opinions are clear:

> We must not force local overseas employees to identify with the Chinese way. What is the Huawei culture? It's like an onion with many layers; one layer is the British culture, another layer is Chinese, and still another is American. So ours is an open and inclusive culture. We shouldn't require things of them based on the Chinese way of thinking, but make the most of their gifts with an open mind in order to enrich our culture.

In 2001, someone in Huawei proposed to set up an association of people with doctoral degrees within the company; it was unequivocally denied by Ren Zhengfei. He said that this would be a retrogressive organization and that if the company was going to set anything up, it should set up an "Open Society."

Openness is easy to talk about, but it's a difficult thing to achieve. China's road to openness over the past three decades has been bumpy. In its first 10 years, openness was not in Huawei's dictionary: the first thing on its agenda was survival. It had to win contracts, gain market share, and struggle like a pack of wolves, each person a hero, engaging in a guerilla approach to the market. This was Huawei's formative period of hunting and gathering—its primitive stage of accumulation—when survival was the most important task. The company had to grow up fast or get slaughtered in the market, where cutthroat competition had forced out more than 400 market players.

As Huawei grew larger, openness became a matter of life and death. The company was born as an underprivileged competitor: It was a private company with no abundant capital, no proud history, no political background, and no technology to call its own, and none of its founders had ever managed any other company before Huawei. They were in a complete and utter void. Against this backdrop, Huawei was given the luxury to maximize the organization's spirit of openness and progress: It simply had nothing to call its own and therefore nothing to lose. It had no choice but to open up, especially in the international market, where a closed mindset would end the game.

Here I would like to include an excerpt from the American botanist Daniel Chamovitz, who made an interesting comment on plants in his book *What a Plant Knows*:

> People have to realize that plants are complex organisms that live rich, sensual lives. [...] But if we realize that all of plant biology arises from the evolutionary constriction of the "rootedness" that keep plants immobile, then we can start to appreciate the very sophisticated biology going on in leaves and flowers. If you think about it, rootedness is a huge evolutionary constraint. It means that plants can't escape a bad environment, can't migrate in the search of food or a mate. So plants had to develop incredibly sensitive and complex sensory mechanisms that would let them survive in ever-changing environments.

This paragraph is a perfect explanation of why Huawei had to open up: It had to survive amid inherent constraints and the ever-changing environment.

Lagging Behind the Times Is an Advantage

In 1992, a group of four people from Huawei, including Ren Zhengfei, Guo Ping, and Zheng Baoyong (executive vice president), went to the United States for the first time. They didn't have a specific purpose but just wanted to experience the outside world. Over the course of 10 days, they covered half of the country. During the trip, Ren Zhengfei caught a bad cold. They were staying in the small inn of a church, eating congee and Chinese pickles every day that were prepared by a local person of Chinese descent. At that time, none of the four could adapt to Western food. Later, on an informal occasion, Ren Zhengfei proposed that "to be international, we have to first develop a Western stomach and learn to eat cheese." This is likely to have come from that trip to the United States.

There is another interesting anecdote from this trip. It happened to be during the presidential election in the United States. Bill Clinton had beat George H.W. Bush and was the new President of the United States.

The world had completely changed around that time with the fall of the Berlin Wall, the newfound sovereignty of multiple Eastern European countries, the collapse of the Soviet Union, and the unification of Germany. For Ren, who had been very sensitive to global political news, the trip presented a rare opportunity to observe the Western world more closely. One day, in the lobby of the hotel, a group of Americans of different ethnic backgrounds were dancing around in excitement while staring at the TV screen, celebrating the newly elected President. In a sense, they were welcoming the arrival of new era in which the United States would dominate the world. Ren had just turned 48 at the time, and fired up by the contagious atmosphere of revelry, he tried to squeeze through the crowd and get closer to the TV. Instead, he got stuck in the middle of the crowd and started dancing along with them, the breast pockets of his suit jacket stuffed with US$50,000 in cash (the entirety of their travel expenses in the United States). Dancing about, his jacket bulging—it was a funny scene, indeed.

Lagging behind the times is also an advantage. Aware that they were laggards, Huawei's people absorbed new information like a sponge. In the 20 years following Ren's first trip abroad, tens of thousands of the company's Chinese employees went out to every corner of the earth. Most people went overseas straight from college: Some of them did not know how to tie a tie, had never eaten Western food before, couldn't speak English fluently, or didn't know a single word of it at all. But they were fearless because they were ignorant and they were humble because they knew they were behind the times. Most of them treated the world as a huge, unwalled campus and forced themselves to learn with an open mind every day, starting with language, customs, food, and clothing, to systems, ideas, technology, and culture.

Lu Ke, the head of Huawei's HR Committee office, was formerly in charge of Huawei's research center in India. He set a strict rule that all communication in every corner of the office had to be in English, and outside of work, Chinese staff had to speak in English as much as possible. This way, their English improved quickly and, at the same time, it brought Chinese staff and Indian staff much closer together. These days, this is standard practice for most of Huawei's branches in nearly 170 countries and regions.

According to a former executive vice president at Huawei, Fei Min, for a long time since its stages of initial development,

> the company was like a mini version of the United States in the aspects of corporate culture, work atmosphere, interpersonal relationships, team work, value evaluation, interdepartmental cooperation, pervasive entrepreneurial drive, and views on product and market. It was all very similar to the United States. This made internal management simple and easy, and at the same time helped the company naturally adapt to global collaboration and communication with overseas clients.

> The United States is the most developed market economy, with a solid foundation across all aspects. For companies, they've generally got better internal and external resources, so it's relatively easier in the US to build an enterprise that lasts for centuries. In China, building a company was almost like building a high-rise in the desert. Growth has been much more difficult for Huawei than its counterparts in Silicon Valley.

Therefore, Huawei would have to learn from others with an open mind. Learning is the only way to grow and develop.

Chen Lifang, a board member at Huawei who's in charge of global public affairs and communications, often represents the company in authoritative forums, where she communicates with politicians, scholars, and media dignitaries from around the world. As Ren Zhengfei puts it, "She deals with all the big shots." Chen has a strong sense of propriety and equilibrium, being both a great listener and an agile learner. She reads widely and extensively and is sensitive to external information. Even for an organization dedicated to learning like Huawei, her ability to absorb knowledge is still quite exceptional.

The 31-year-old CFO of Huawei's chip company, Ye Xiaowen, has a double bachelor's degree from Jinan University and a double master's from the Chinese University of Hong Kong. During his eight years at Huawei, he has bought an average of 100 books every year, reading two every week. He commented:

> Initially the gap between Huawei and Western companies was no less than the gap between 0 and 100. The only way we could

catch up was to learn hungrily. Afterwards, when we got closer to our competitors, our boss advocated self-criticism across the whole company to prevent arrogance and closed-mindedness. This culture of self-criticism was like carving out a drainage basin, so that the waters from higher points in the business terrain—companies in the West—would eventually flow down and settle in the lowlands at Huawei.

Huawei is undoubtedly one of the companies that benefited most from the era of the Internet and globalization. I visited Japan twice in the first half of 2014, where I had the privilege of meeting a number of renowned Japanese entrepreneurs, including Kazuo Inamori. The president of a large century-old Japanese company said, "Huawei is a global company with a global platform. Japanese enterprises are very strong with vertical integration. We are confident that we can sell products to the world by cooperating with Huawei."

The Americans Scrubbed Huawei Clean

The year 2003 was a low point in Huawei's history, but also a major turning point. The Western New Year had just passed and the Chinese Lunar New Year was just around the corner. It was January 22. Cisco, the global communications giant, caught Huawei off guard with a lawsuit for intellectual property-rights infringement. The indictment was over 70 pages long and it covered nearly every type of intellectual property law in existence. Litigation was to take place in the Marshall Division of the United States District Court for the Eastern District of Texas. Analyzing why Cisco would have chosen this location for prosecution, an US lawyer said, "This courthouse is known for quick judgments on intellectually property suits, and for its severe penalties. The folk in Marshall are fairly conservative, you could say xenophobic."

Almost immediately, the Western media—and even a number of Chinese media outlets—began to smear Huawei left and right, painting it as a "hideous little robber" or a "technology thief," and suggesting that Huawei was in over its head. One of Cisco's senior executives in the Asia-Pacific region, an overseas Chinese, said in a public statement: "This time we need to make Huawei go bankrupt."

Cisco had been planning for the lawsuit for over a year. Starting with the US government and then relevant departments in the Chinese government, on to preparing their team of lawyers and public opinion, and then choosing the time and place of litigation, it was all an act of painstaking design. Huawei was not in the least bit prepared for what just hit it.

Huawei's senior executives were hastily thrown into the foray. Americans were the ones who had taught Huawei the meaning of international rules and they were also the ones who taught Huawei the art of imitation—in this case, adopting an international approach in countering their US rivals in court. As the legal battle proceeded, some recommended that Huawei position itself nationalistically as "the defenders of Chinese enterprises" in order to win local media and government support, but the company's decision makers rejected this approach. They believed that because this was an international lawsuit and the trial was being held in one of the United States' most conservative states, besieging the besiegers would only complicate matters; Huawei would have to bravely answer the charges in court, because only by boldly facing their challenger could they broker peace. They believed that, even if they lost the suit, although it would be a painful experience, it wouldn't spell the end of the company. On the contrary, were they to play the nationalism card in the company's defense, even if it survived the crisis this time around, it would be the end of Huawei's development in the international market.

And the fact of the matter was that, given the global nature of the IT industry, it wasn't possible for Huawei to be a fully "Chinese" company. From day one, Huawei was facing world-class competitors—not only did it have to compete with them, it had to learn from them as well. Of course, learning from them was the most important part. As one of Huawei's senior executives once put it, "For the past 20-odd years, we really came out with the better end of the bargain. After all, we had someone to lead the way forward. Alcatel, Ericsson, Nokia, Cisco—they all showed us the ropes." And what ropes did they show Huawei? The direction that products and markets were heading in, management models, and so on. But most importantly, they got Huawei up to snuff on international business regulations.

In and around the 1990s, competition for the communications market right on China's doorstep was the earliest and most large-scale form of international competition in China's commercial history. It set the precedent for global competition among all the Chinese industries in the years to come, providing them with relevant experience while also cultivating a group of Chinese companies like Huawei and ZTE that were capable of competing in the global market. They not only became the strong opponents of Western companies in China, they also became the forerunners in the global playing field. So in those days, were Huawei to position itself as a "national enterprise" in China and not a global company, had it not shouted its international ambitions from the rooftops like its competitors, then it simply wouldn't be the Huawei it is today: a global industry leader. And history shows this to be the case: Most of the companies that position themselves as a national enterprise in China have a difficult time expanding beyond their own borders.

Business has no nationality and markets know no boundaries. How many Western companies in the Fortune 500 do you see waving their national flags around? Their ambitions are global and yet they make a core contribution to the strength of the countries where they're brought into being. Huawei leadership believes:

> Globalization is inevitable. We need to boldly open ourselves up and proactively compete with the West, learning how to manage ourselves better along the way. For more than 10 years we have never once referred to ourselves as a national industry here in China because we are global. If we close the doors to the rest of the world and rely only on ourselves to survive, the moment we open up the doors again, we'll collapse at the first touch. At the same time we need to work hard to support our globalization process with our own products.

Ren Zhengfei has never been one to shy away from decisions and he firmly believes that a well-handled crisis is likely to become an important opportunity. At great expense, Huawei hired strong legal counsel in the United States and met Cisco head-to-head along two tracks: the lawsuit itself and in the media. The dramatic ups and downs, twists and turns were no less thrilling than a war novel. A year later, on July 28, 2004, the dust finally settled on the lawsuit that 3Com's global CEO, Bruce Claflin, called an "entertaining theatrical performance." The suit ended with a mutual settlement: each company left to its own devices and each responsible for its own attorney's fees. There were no apologies, certainly no indemnities, and it was ruled that Cisco could never bring the same claims against Huawei in the future.

Beginning in 2003, Huawei's products made their way throughout the entire European continent, including Western and Northern Europe, and then further into Japan, South America, and North America. As of 2010, 70 percent of Huawei's sales revenue comes from international markets.

The fact that Cisco couldn't discredit it just so happens to be Huawei's best word-of-mouth recommendation: "Have a look—Huawei's clean. The Americans scrubbed it right up. It is a capable company, and even a company like Cisco is quaking in its boots."

An executive in Huawei's enterprise network business group remarked with a sigh, "Huawei's enterprise network products are far better known in America than in any other region…."

The unbearable arrogance of Western companies like Cisco played a major role in arousing the competitiveness of Chinese enterprises. One night in 2003, dozens of R&D staff in Huawei's Beijing Research Institute, all around 30 years old, got a little tipsy after a group dinner and started shouting at the top of their lungs: "We don't believe we can lose to Cisco. We just don't believe it!" A casual dinner had turned into a pep rally.

Ten years later in 2012, it was this same research institute, now with nearly 10,000 R&D staff, that developed the 400Gbit/s router 12 months ahead of Cisco, having already achieved commercial application in many countries worldwide. Twenty years ago, Cisco was an advocate of IP technology. Today, however, Huawei's Beijing Research Institute has become one of the leading IP R&D centers in the world.

Attempting to Dominate the World Market Will Spell the End of Huawei

Throughout the Cisco lawsuit catastrophe and in the process of turning it around, Huawei learned more than ever. The two-year litigation placed Huawei under the most stringent scrutiny. Ever since then, every inch of the company has been pored over as if through a giant magnifying glass. As it progresses forward into the global market, especially in Europe and the United States, Huawei has been repeatedly subject to the "interrogations" of customers and even governmental organizations alike.

A Chinese official once asked Ren Zhengfei, "Do you have any experience in the international market that you can share with other Chinese companies?"

Ren Zhengfei answered:

> The key is to abide by the law. We must observe the laws of the countries where we operate and the conventions of the United Nations. In particular, we'd better treat the domestic laws of the US as international law, because the country is so powerful that it can come after you on the basis of its own domestic law. On the other hand, China's legal system is not yet complete, and enforcement tends to be a bit arbitrary and random. As a result, solid governance and self-discipline are missing in some Chinese companies. Many think they can easily acclimate and succeed in the international market, but they end up getting themselves into trouble.

Ren Zhengfei often cautions his leadership teams that Huawei must not subject itself to the adjudication of its competitors. In business, circumstances change, and the concern is that leaders become too stuck in their ways. For the company not to be labeled as a "mysterious black widow" or a "brash gladiator," it has to stay open across all aspects of its business. And even if the company already thinks it's open and transparent, because openness itself has enabled the company to grow, if Huawei is still accused of being "closed off" or "unconventional," then it goes to show that the company isn't as open as it ought to be. To illustrate his point, Ren Zhengfei provided an analogy: Imagine you are invited to a friend's home. Now imagine how much they would despise you if you kicked off your shoes and began scratching away at your toes in the middle of the living room. Huawei must not behave rudely. We must prove, with an even more open-minded attitude, that we are a company that plays by international rules.

In the years surrounding 2010, Huawei managed to go from a backward company to a leading industry player. Will it manage to hold on to the strengths begot of its backwardness, its humility, and the advantages of its open culture? Should it hold on to these at all? Can it escape the historic trap where strength gives birth to tyranny and magnitude to pride? In a meeting with his senior management team, Ren Zhengfei warned them in an extremely forbidding way:

> Anyone in a position of strength needs to strike a balance. We can be as strong as physically possible, but can we last if we have no friends? The answer is no. So why should we go out and knock others down, and take on the world alone? Genghis Khan and Adolf Hitler had wanted to eliminate others, to conquer the world, but in the end they were destroyed. Huawei will surely be destroyed if we set our sights on dominating the market. We'd do far better to unite our strengths, and work together. We shouldn't have such narrow-minded view of business: Our relationship with others should be a blend of competition and cooperation. As long as we benefit in the process, that's perfectly fine.

He further commented:

> When we work with others, we can't act like black widows. Black widows are spiders found in Latin America. After mating, the female spider eats its male partner in order to provide nutrition for their offspring. In the past, after forging partnerships with other companies, Huawei used to eat them up or toss them aside after just a few years. We're already strong enough as is. We must be more open, modest, and look a deeper look at the problems that lie before us. We mustn't be narrow-minded, focusing on petty victories, lest we become a bunch of fearmongers and tyrants. We need to find a more effective model of collaboration to ensure mutual success. Our R&D teams are quite open, but they need to open up further, both internally and externally.

> Huawei has grown strong, and not everyone out there loves us. There are those who hate us because, in effect, we may have starved them of their business. We need to change this situation; we need to open up, collaborate, and ensure shared success,

because true success can't be built on a pile of bones. We may have turned many friends into enemies over the past 20 years. For the next 20 years, we will turn them back into our friends. For when the value chain is built on friendship, then success is the only way forward.

Huawei's View on Openness: Be Open, Not Allied

Acquiring Ideas and "Paying the Toll"

On January 27, 2009, in an announcement about the number of international applications filed under the Patent Cooperation Treaty (PCT) in the year 2008, the World Intellectual Property Organization (WIPO) published the following news on its website: "For the first time, a Chinese company topped the list of PCT applicants in 2008—Huawei Technologies Co. Ltd, a major international telecommunications company based in Shenzhen, filed 1,737 PCT applications in 2008."

Huawei's rise to number one in the world led to exultant media coverage in China at a time when independent innovation was hailed as a national priority. Meanwhile, a small minority of international media was commending this accomplishment, but they remained cautious and doubtful.

Within Huawei itself, the announcement was met with nothing but silence. There were no celebrations within the company. On the contrary, at an executive management team meeting, some senior executives emphasized the downside of such a large number of patents: "Having the highest number of patent applications does not ensure quality. Has it earned us any royalties?"

Following a decade of quiet transformation, Huawei has developed a unique culture of composure, rationality, self-restraint, and self-criticism. When the company made its way into the list of Fortune 500 companies for the first time, an executive walked into the meeting room and said, "I have some bad news. We have become a Fortune 500 company." No one applauded or proposed a celebration. In fact, Huawei had been trying to postpone its entry into the list.

Huawei's leaders were very sober: Although Huawei was possessed of great technological strength, in the global information-technology arena it was still a latecomer, a key player and nothing more.

Up to now, Huawei has not produced a single original invention. What it has done is enhance the functionality or features of products developed by Western companies or made them more integrated. In other words, Huawei's technological achievements have been primarily on the engineering side. There is still a large gap between Huawei and its competitors, who have been accumulating technological expertise for decades or even centuries. To compensate for its lack of core technology, Huawei has acquired or licensed the technology it needs and this has given it access to the international market, allowing it to survive in a competitive marketplace. At the same time, this approach costs much less than reinventing the wheel, buying peace with Western companies by paying for the use of their technology.

In 2010, Huawei paid US$222 million in patent royalties to its Western counterparts and US$300 million in 2013. In total, the amount of IP royalties that Huawei paid to Qualcomm alone is close to US$1 billion, which undoubtedly is the largest amount that any Chinese company has ever paid. As huge as these royalties are, they have earned Huawei billions' worth of contracts. Once, when an R&D executive reported their engineering achievements, Ren replied with a smile: "Come brag to me when we stop paying US$200 million a year."

In the past, Fei Min was in charge of Huawei's R&D. He held that the industry is a supply-chain system, in which for the longest time in the past, Huawei's innovation was to get close to the market and understand client needs. Leveraging the most advanced chips, basic software, and operating systems developed by other companies, Huawei uses open architecture to integrate resources, and then develops solutions and products for systems and networks that better meet their customers' needs.

In order to remain focused on customer needs, the company has to maintain an open culture of research and development. Openness isn't even a choice, really. An executive explained with a metaphor:

> Huawei has managed to carry its banner from the base of a mountain, up to the middle, and at last to the top. But looking down from the peak, it realizes that each step of the mountain pass has been surrounded by enemies. Virtually all essential patents in the telecom industry belong to Western companies. After more than 10 years of technological accumulation, Huawei has been caught in an unpleasant situation: Yes, it stands at the top, but it has been trapped up there, barricaded by enemies at the base of the mountain and all along its slope.

> What can Huawei do? It will starve if it remains on the mountain top. But if it makes its descent, it can't steal or fight its way down—the odds of survival aren't great. The only way is to buy "passes" or "pay the toll." In other words, it has to pay royalties for patented technologies or engage in patent exchange.

Since the early 21st century, Huawei has encountered more than a dozen patent disputes or lawsuits each year. From time to time, Ren Zhengfei receives letters from CEOs of Western companies claiming they will sue Huawei unless he pays them a hefty price for alleged patent infringements. While the Cisco case was a majorly stressful event back then, such patent lawsuits have become routine. Huawei itself has also taken legal action against other companies over IPR (intellectual property rights) infringements. This is a common occurrence in the global telecom industry, in which the arena may seem overcast with intense hostility at times, but then the dark clouds suddenly disperse as quickly as they appeared. Almost all lawsuits in the telecom industry are ultimately settled peacefully out of court. Why? "That's the rule of law in this world," as Huawei's rotating CEO Guo Ping puts it.

He went on to say, "If you hope to enter the 'international club,' you have to pay the toll and play according to international rules in order to leverage others' strengths. How can you expect to use the proprietary technologies of other companies for free? They have acquired their intellectual property at a high price. On the flip side, if they want to use our technology, they have to pay us too. If need be, we can exchange patents. It's only fair."

After 2008, Huawei has made peace with its Western counterparts, although there are still some flare-ups over patents from time to time. This improved relationship is the result of openness as well as a balance of power. As Huawei executives predict, in the coming 3–5 years, there will be a world war over patents, and for the most part, China will only be able to engage in a defensive role. Huawei shouldn't let its guard down.

The good news is that Huawei has gained enough strength and experience through its battles with the Western companies. Most importantly, Huawei has developed an open strategy and an open mindset, which are the secret weapons behind the company's ongoing success.

Alliances Are Like Ropes That Tie You Down

As the idiom goes, success depends on your opponents. The Cisco case was a milestone for both companies—Cisco has become increasingly defensive and closed, while Huawei was prompted to become more open and progressive.

Huawei has since reflected on its strategic positioning: It is clearly not realistic to depend solely on itself, although it has become stronger after a decade of development and is able to weather most storms. During its first 10 years of existence, Huawei behaved like a noisy, furious, fearless, and aggressive wolf that seized on every chance to grow. But during its second 10 years, it became quieter and much more restrained as it transformed and expanded in peace. Now, Huawei has emerged as a world-class corporation, a gigantic ship that can sail through turbulent waters. It has turned the tables around, or has at least grown to be on par with its Western counterparts. Therefore, Huawei can afford to open up its doors wider and more boldly. Now Huawei has given up the tradition of fighting alone; its new principles are profit sharing, mutual benefit, compromise, and cooperation—but not alliance.

Ren Zhengfei believes that, in this modern world, with new technology popping up everywhere and the market more fickle than ever, Huawei can't afford to do everything from scratch itself and it does not possess the capacity to dominate the market alone. Neither does any top player in the West. In this context, companies have to combine forces and foster partnerships with their competitors in order to grow strong.

In 2005, Huawei's leadership team emphasized partnerships with competitors as a strategic priority for future development. They said:

> A few years ago, we proposed to give land for peace, and to build up partnerships with our counterparts to complement each other's strengths and create more value for our customers. Now, we have been recognized by more people, and our competitors have come to treat us as their friends. While we compete to provide better customer service, we work together to lower development costs. This change in the paradigm has fueled our growth momentum and will reshape the trajectory of Huawei's development.

In recent years, Huawei has forged worldwide partnerships for technology and market development. The company has set up joint labs with Texas Instruments, Motorola, IBM, Intel, and Lucent Technologies and has established research institutes in India, the United Kingdom, France, Sweden, Italy and Russia. Among Huawei's 16 research institutes, eight are outside of China. Meanwhile, 31 joint innovation centers were set up with clients to jointly explore their future needs; they also established various forms of innovation partnerships with nearly 100 colleges worldwide, comprising a highly open approach to research and development.

Of course, partnership is not equal to alliance. Any alliance, in essence, is a closed-ended system that runs counter to openness. If it enters into an alliance, Huawei would confine itself to a different type of closed system or even become the prey of its allies. In history, numerous alliances between countries or between companies have ended in tragedy. For over 20 years, Huawei has consistently avoided the formation of a so-called "united front" or an exclusive regime with any industry giants. For Huawei leaders, a one-sided "strategic alliance" can only lead to more enemies, which is not a good thing for Huawei.

The second year after the Cisco case ended, Microsoft CEO Steve Ballmer visited Huawei headquarters and met with Ren Zhengfei. Without much preamble, he directly requested Huawei not to join the antimonopoly movement against Microsoft. Ren said that preventing monopolies is the government's responsibility. It drives innovation, benefiting society and consumers alike. But it's not within Huawei's purview, nor does Huawei have the capability to prevent the monopolies of other companies. Ren Zhengfei, of course, knew very well that Huawei had driven Cisco back with antitrust appeals and open competition was critical to its survival. But Huawei would not proactively involve itself in any antimonopoly movement. He said:

> Why can't Huawei stay under several umbrellas at once? We can hold the Microsoft umbrella with our left hand, and the umbrella of another industry giant on the right—they can sell for higher prices, I'll sell for lower, and make a ton of money doing it. If I dropped one of the umbrellas, the sun would bake me alive, and my sweat would pour down in rivers, nourishing the grass below. This grass would grow up and start encroaching on our territory with even lower prices. There's no way I'd do something quite so thankless as that.

"Drawing Energy from the Universe over a Cup of Coffee"

One afternoon in October 2011, Ren Zhengfei met with Eric Schmidt, executive chairman of Google, at Google's head office in Beijing.

Schmidt greeted Ren, saying, "Huawei's story is the most successful in China. You're changing the world. You're a great magician."

Ren Zhengfei replied,

> We are still in the 18th or 19th century in the West, the cost-and-quality days of the Industrial Revolution. We entered this industry because we're a bunch of fools. We've relied on cheap labor for two decades, but as the cost of labor rises in China, we will gradually lose this advantage. In several years, we expect our

sales to reach US$60 billion, but we can't count on continuous expansion alone. We've survived by driving up scale to lower the variable cost per unit, but this won't work forever.

In response, Eric Schmidt said, "All my life I've dreamed of connecting the entire world. Google has developed close partnerships with many companies. We're convinced that Huawei will help lead the future."

"Unlike the movie *2012*, the flood will not subside," Ren replied. "It will only get heavier. Of course, when I say 'flood,' I'm referring to a flood of data. Google provides the water, and we provide the pipelines. We firmly believe that this world is full of enormous opportunities."

Eric Schmidt couldn't agree more.

Ren Zhengfei, now 70 years old, always has his mind on overdrive, is always open to new ideas. He is interested in books on politics, economy, society, humanity, literature, and the arts. He reads history books the most, fiction and management books the least. He explained: "Fiction is just invented stories; they are too far from reality. Management books are produced by professors behind closed doors. They restrict the imagination. Corporate management can't be boiled down to just a few tenets."

Communicating with all sorts of people is the most important part of his job. Over the past two decades, he has visited almost every country in the world, from the most developed to the least. From countless conversations with hundreds of politicians, business leaders, scholars, competitors, scientists, artists, and monks, he has inadvertently broadened his world view. Perhaps Ren Zhengfei is the best informed business leader in China and China's most outstanding business thinker.

Ren Zhengfei encourages and requires his leadership teams and the company's scientists to bounce thoughts around with influential people in the world over a cup of coffee. The global village, in some sense, is an open college full of information and knowledge.

According to Ren, coffee breaks are an opportunity for people to hang out and exchange ideas with others, because you never know who you might run into. And some might ask: Why not a cup of tea? He believes that tea has its own Eastern charm while coffee is a symbol of global culture. Ren Zhengfei argues that Huawei is part of this tempestuous world and must stand firm at the forefront of its trends and culture.

Often mocking himself as a country bumpkin, Ren Zhengfei is honest and candid. He never beats around the bush. After a meeting with Ren, Texas governor Rick Perry said, "Mr. Ren is a very interesting man. He is as straightforward as a Texan. He won't just say what you want to hear." He went on to assess the company, saying that Huawei represents high standards.

Unlike Ren Zhengfei, Huawei's chairwoman Sun Yafang is more elegant. As a Harvard Business School graduate, she speaks fluent English and behaves with propriety. On account of this, Ren Zhengfei has nicknamed her "Huawei's business card." Unlike Ren Zhengfei, who easily dominates the conversation with his charisma, causing "creative friction" from time to time, Sun Yafang is more patient and skillful in communicating with the big potatoes.

Hu Houkun, Huawei's deputy chairman and rotating CEO, who was once in charge of global sales, exudes a more peaceful strength. At various global summit meetings, such as the World Economic Forum in Davos, Hu meets with world leaders in a very casual manner, chatting in the lobby over a cup of coffee with Sun Yafang, Guo Ping, Xu Wenwei,[2] and Chen Lifang.[3] In recent years, as Huawei's ambitions in the US market have met with repeated setbacks, Hu and Guo visit the country frequently to meet with politicians, business leaders, think tanks, and public relations agencies, having candid conversations and answering their inquiries without reserve.

Guo Ping, another deputy chairman and one of Huawei's rotating CEOs, has also earned a reputation in the US IT and legal communities for his central role in combating lawsuits filed against Huawei in the United States, including the Cisco case and a subsequent legal dispute with Motorola.

---

[2] Also known as William Xu, Xu Wenwei is the president of Huawei's Strategy Marketing department.
[3] Chen Lifang, mentioned briefly in previous chapters, is the president of Huawei's Public Affairs and Communications department.

As one US citizen put it, the West is not a monolith. There are divided opinions between the European Union and the United States, and even between the United Kingdom and the United States. (Author's note: In fact, Huawei's investments and trade with the United Kingdom have always gone quite smoothly.) The US companies do not speak with the same voice, either. So Huawei has made the right choice in communicating openly and extensively. The US market will accept Huawei one day: It is only a matter of time.

Huawei's people believe similarly that the world isn't dark all over: When it's dark in the East, it's light in the West, and when it grows dark in the West, there's still the South. Sooner or later, Huawei will greet a world that all shines brightly under the same sun. In the meantime, what Huawei needs is strategic patience. It just so happens that Huawei's development throughout all these years has been a carefully balanced dance of knowing when to be patient and when to move forward.

For Huawei, according to Ren Zhengfei, it is important that everyone opens up their minds and starts considering the world at large. He said, "Don't just think 'me, me, me,' but think 'the world, the world, the world.' You need to think about how you can make your own contribution to the world and you'll feel proud of yourself when you do. That's all it takes."

Openness Is an Art of Thought

Visibility Attracts Danger

People these days seem to be afraid of quietude. Huawei was born among the grassroots and to many people, its rise has been astonishing. The company awed its Western counterparts in the first 10 years, and in little more than 20 years, it has soared to the number-one spot in the world. At first, few believed that this was the result of market competition, and some even invented stories about the company's background. Now there are still quite a few people, both inside and outside the industry, who argue that Huawei's success is credited to unusual, non-market factors.

And this does raise the question that, if there's no funny business, why doesn't Huawei go public?

"How is it possible?" some wonder. How could a Chinese company become so successful in less than 30 years' time and pose such an immense threat to a whole bunch of Western companies? What's so special about Huawei? Why isn't it some other company?

Most Chinese entrepreneurs are caught in a dilemma: If you're too vocal, you get criticized as "forgetting your place," whereas if you're too quiet, people think you're hiding things. It's as if there are microscopes and circus mirrors swinging around on both sides, warping the situation until you fall over and die—the only chance you'll ever get at proper rest and respite.

For quite some time, Huawei has had an almost morbid fear of the media. Ren Zhengfei has seen so many tragic heroes fall in this Age of the Internet, when a drop of spit can set off a tsunami. Every year in the 1990s, the top 10 entrepreneurs all met an untimely demise. And of the top 10 named "most wealthy" in the first decade of the 20th century, how many are still around? Among the business celebrities who danced among the headlines, those flashy men and women of the times who stood in the spotlight, how many are still on stage?

Since the reform and opening-up program started in China, a number of business adventurers have reached center stage, including Nian Guangjiu, who made a fortune selling fried sunflower seeds; Bu Xinsheng, who shattered the "Iron Bowl" of state-owned enterprises, on which the employees of his factory had been too comfortably fed; Ma Shengli, who was the first businessman in China to be granted rights to run a company through a lease contract; Lu Guanqiu, who grew from a farmer into a large corporation owner; Chu Shijian, who championed the reform of state-owned enterprises; Ni Runfeng, who turned Changhong into a leading producer of televisions; Mou Qizhong, who trod the line between politics and commerce; Shi Yuzhu, who carved his spot in the market through wave after wave of advertisements; Zhang Ruimin, who created the legendary Haier model; and Liu Chuanzhi, who guided Lenovo's transformation from a technology-led company to a sales-led conglomerate.

Sadly, many of these adventurers failed to resist temptation. When each company was at its most critical impasse, they threw themselves into the warm, cloying embrace of the media. As a result, many stars became falling stars, almost overnight. This is a tragic phenomenon for those in the Chinese business community and a miserable page in the history of Chinese business.

Climbing uphill is tough, while falling downhill is much easier and much faster. The most tragic part of the story is when businessmen opt out of their original role and, enticed by the lure of the limelight, put on red dancing shoes and a face full of makeup, and enter the stage as actors in a new role. They may not realize that the same people who cheer the loudest are often the first ones to denounce them and eventually help dig their grave. And when they lose their luster, they find themselves alone in a quiet corner, forgotten.

As a Chinese saying goes, "Look and see a tall building; look again and see it's fallen." During the 30 years of China's reform and opening up, numerous business stars rose to take center stage, only to quickly fade away into a short narrative rich with tragicomic elements. On one day, they might have been pointing towards the vast expanse of the sky, imagining all the possibilities as they gazed off into boundless distance; the next day, they might just have found the ground littered in fallen leaves, their prospects cold and desolate. And why is this? The reason is very simple: Exposing yourself attracts danger.

The mission of a merchant is to make profit; seeking fame is anathema. In this diverse world, everyone or every trade plays a definite role and the title of "star" is reserved for performers, politicians, and their ilk. Entrepreneurs and businessmen, on the other hand, should keep themselves as far away from the stage as possible. No one can attain both profit and fame at the same time, so businessmen should race away from fame just as politicians shouldn't race towards profits. Over the past three decades, too many Chinese business leaders, in the face of a little success or ill-formed glory, couldn't help but jump into the limelight for a taste of victory, only to find that they were no different from moths jumping into the flames. Their ends are tragic, of course. This sweet taste of fame is transient and, worse still, their companies often fall along with them.

On the contrary, those who have held themselves back from the carnival are those who are most likely to survive and become true leaders. Treading quietly for the time being, they are the ones who may very well become the most lasting stars in the commercial Milky Way, building world-class business empires that are lit by the failing light of more fickle, falling stars.

1998–2008: A Decade of Being Open and Closed

Opinions concerning Huawei's openness differ greatly. The company's management believes Huawei is open. As Ren Zhengfei said:

> Huawei must embrace globalization and avoid the narrow mindset of nationalism and protectionism. Therefore, we have tried from the very beginning to create an open culture. Through competition with Western companies, Huawei has learned to compete, and has improved its technology and management. Only by tossing aside our narrow sense of nationalistic pride, our narrow sense of Huawei pride, and our narrow sense of brand awareness, can we become a truly global, professional, and mature company.

The decade from 1998 to 2008 was Huawei's most open period, the decade where Huawei learned the most from Western companies. During this period, Huawei established partnerships with more than 400 telecom operators worldwide. This was also when Huawei began working with more than a dozen consulting firms from the United States, Europe, and Japan on IPD and integrated supply chain (ISC) transformation to develop a management platform better suited for the international market. Huawei's former vice president of procurement was German and it's said that "during the two years he was in charge, procurement went from some backwater affair to a fully modern procurement system. Even in the hardest years after the IT bubble burst, we were able to save more than US$2 billion a year. Our procurement system performs well above international standards."

Huawei's global cyber security officer (GCSO), John Suffolk, was the former chief information officer and chief information security officer

(CIO and CISO) in the UK government. After working at Huawei for two years, he had the following impressions:

> The surface-level understanding that most people have of Huawei isn't the real Huawei. It's a very professional and open organization with a strong sense of integrity.

> Realizing customer value has helped Huawei perform swimmingly in the UK, and in terms of value systems and working styles, it's enabled Huawei and its UK customers to work together quite well.

> Huawei's processes are very systematic. I put a great deal of stock in processes. They can be very dull, but they're extremely important. …Huawei's culture stresses that patience and hard and solve any problem.

> Huawei isn't looking at how to make the most money. All things begin with exploring how to satisfy customer needs. This is the primary reason behind Huawei's success.

Huawei now employs some 40,000 non-Chinese nationals, and the presidents and chairmen of Huawei's subsidiaries in the United States, Europe, and Australia are mostly foreigners. With its international competitors, the relationship is one of competition as well as partnership.

So is this enough to prove that Huawei is an open company? Perhaps not. The problem lies with Huawei's media strategy, or its strategy of openness.

Openness is an art of thought. It takes great deliberation for an organization to decide when it should open its windows and when it should open the door. When should it open both? To whom? And how wide?

Huawei's leadership has drawn a wealth of inspiration about corporate governance from the progressive nature of China's opening-up program. Essentially, the company had to open up across the board; this was the megatrend of the 1980s. But the company could still control how and when it opened up and at what pace. It could open up at different levels, both externally and internally, according to its own strengths and maturity.

In 1998, Huawei started its adaptive reform program. The top management of the company understood that this radical process of transformation to Westernize the company would surely trigger cultural clashes or even extensive resistance. If managed poorly, the reform program would fall through and the company might very well collapse. After all, this was a large-scale organizational reform that would substitute processes for human effort, automation for personal will, and cold rules for interpersonal relationships. It was an unprecedented initiative among Chinese organizations, and even now there aren't many successful cases.

Because of this, it was especially important to rule out any internal or external interference, especially from the media. According to Xu Zhijun, Huawei's deputy chairman and rotating CEO, this was a critical program that would decide the fate of the company, so the company had to be left alone to make its own decisions. Moreover, since internal and external environments were already riddled with uncertainty, Huawei's chance of success would be cut in half were the media allowed to meddle in the process. Fortunately, Huawei was able to remain level-headed and shy away from the media for nearly a decade. Had it not done so, the company might have crumbled apart long ago. Huawei was able to fend off the media because it was and still is a private company without any obligation to disclose information. Of course, the company is different from what it was 10 years ago. Now its formative transformation is essentially complete. Compared to its developmental years, it is much better equipped to survive.

Nevertheless, the media and general public are of a different opinion. Huawei is portrayed as a closed and mysterious company. Few seem to recognize Huawei's open approach to market and technology development as well as the increased openness of its operations and management.

At the other end of the spectrum, there are companies in China that are very welcoming towards the media but, in fact, they lack the nerve to open up their R&D, market, and organization in any real sense. This contrast deserves attention.

Huawei's estrangement from the media began in 1998, when its sales totaled CNY8.9 billion, the highest among the four major Chinese telecom-equipment companies. Huawei had reached its first peak: number one in China. But at that moment, Ren Zhengfei and his company were not thrilled or proud; instead, they felt more lonely, anxious, lost, and terrified than they ever had before.

They were wondering what Huawei's next step would be. Huawei had to plan its future and they were aware that they had to undergo a metamorphic organizational and cultural change. Since then, Huawei had opted for silence, shunning interviews and forums and dodging awards like an ostrich hiding its head in the sand.

This silence lasted for an entire decade. According to Ren Zhengfei, the company had kept silent because he, his team, and his organization had to focus on their own transformation and development and not allow past success to fetter their progress.

Without external disruptions, Huawei was able to move forward at full force.

The 10 years from 1998 to 2008 were quiet, and during this time, Huawei and Ren Zhengfei achieved a historic transformation. Huawei did it because it had remained reasonable, quiet, firm, and persistent. But it in the process, it met with a real problem, a true crisis this time.

Open Up Some More—"The Sky Won't Fall"

In a sense, Huawei's history is a history of crisis management. Since its founding, the company has faced many intense predicaments, including crises of capital, talent, and management in its early years, besiegement and obstruction from its competitors, institutional conflicts with the Chinese market system, as well as emergencies that seemed to occur on almost a monthly basis. During the past 27 years, Huawei has suffered a great deal, but in the end, it has survived and grown up. It is just like a tree. The tree takes root in the earth, but this "rootedness" is a huge evolutionary constraint. The tree, therefore, cannot grow unless it has a strong will to weather all manner of storms, absorbing nutrients from the earth to build out its roots, stretching its canopy upwards and outwards amid relentless change and adaptation.

On May 28, 2006, yet another crisis hit. Hu Xinyu, a 25-year-old employee, died from viral encephalitis. The media lashed out, denouncing Huawei as a sweatshop that had worked its employee to death. They went further to criticize Huawei's mattress culture, the fighting spirit behind the company's early survival and growth. On the Internet, especially on some BBS (bulletin board system) sites like *tianya.cn*, there were a flood of bitter messages attacking Huawei, arguing that Huawei was a closed and mysterious paramilitary organization. Most of the messages were apparently posted by Huawei's own employees.

At first, Ren Zhengfei, Sun Yafang, and other Huawei executives were enraged and bewildered. They gave instructions to find out who had posted the messages and wanted to hold the media accountable for their "false reports." To be sure, Huawei had never faced this type of crisis before.

Objectively speaking, Huawei was an open company and Ren Zhengfei was a highly enlightened entrepreneur. Ever since its establishment, Ren had firmly held the belief that "openness is critical to the fate of the company" and "the company will die if it is not open." In fact, Huawei had always been open in terms of its corporate management, R&D, and market penetration, and had taken bold steps to Westernize its organizational structure. To some extent, Huawei differed from its peers because it was more open than any other Chinese company.

The reality, however, is that Huawei was still not open enough. Its organizational culture, in particular, was closed and rigid.

In the history of corporate management, military concepts have played a significant role and have found their way into modern organizations that have tried to improve their structure and efficiency. Ren Zhengfei loves the book *The Whiz Kids: The Founding Fathers of American Business*, which includes the stories of 10 veterans who reformed the Ford Motor Company and brought the company around. This is a very good example. A former soldier himself, Ren had transplanted military tenets into the corporate culture of Huawei, including discipline, order, obedience, aggressiveness, unwavering courage, uniform will, and team spirit. These had been proudly labeled as the organizational characters of Huawei.

At a casual meeting, an employee asked Ren Zhengfei, "What is the most important lesson you learned from your military experience?" Ren Zhengfei replied, "Obedience." Obedience is a natural and unquestioned obligation of soldiers, but the problem is whether it's the same for corporate employees.

The answer is both yes and no. It is true that modern corporate management is inspired by rules of the military. A business organization with a clear goal must have definite authority to hold everyone together and lead them towards that goal. And it must have a definite hierarchy where everyone does their own job strictly according to the rules; otherwise it may not survive in the competitive business world. This is especially true in the standardized assembly line-based manufacturing sector. It is also true for a start-up company. It's apparent that Huawei had acquired the organizational power and competitiveness of its first 10 years by virtue of an absolute authority that led the company through one crisis after another. It had grown on the basis of a definitively designed hierarchy.

After 2005, however, internal and external circumstances changed. First, Huawei started to employ a new generation of intellectual workers: people born in and after the 1980s, or the Internet generation. They were bound in personality: They were rebellious but they would follow the crowd; they were independent but they would rather depend on others; they were open but they would often shut themselves up in their own world. This is the "me" generation. Would traditional management concepts remain valid?

The ripples caused by the death of Hu Xinyu revealed that Huawei was faced with a dual crisis: one both cultural and systemic.

Success is sometimes very fragile and trouble always comes in waves. If the death of Hu Xinyu was an isolated case and if the media had wronged Huawei and Huawei's management, then the suicides of several Generation-Y employees to follow certainly deserved their serious attention.

In a way, crises had forced Huawei open. From 2005 to 2008, Huawei went through a process of self-criticism and confrontation regarding its organizational culture. Questions were raised and debated in light of a more transparent media environment and a changing demographic mix of the workforce. Should the company change its corporate culture? Should the company become softer? Among the key elements of the corporate culture, which should be changed and which should be retained? Should Huawei create a more democratic and liberal atmosphere?

Ren Zhengfei said, "The times have changed. It is time for Huawei to change. We must become internationalized and more open. Why should we avoid disagreement? It's not like the sky is falling."

Once, an employee criticized a regional executive, who later asked the forum moderator for the corporate ID of the employee. When he found out about this, Ren Zhengfei said, "Give them my employee ID if that ever happens again."

Ren Zhengfei explained:

> The more we discuss it, the clearer the truth will be. We should embrace different opinions, because I believe most employees are committed to the company. This is a strategic reserve. We thought at first that this approach might be risky, but now we realize we were wrong. Moreover, any agreement our employees can reach make things easier for management. We can hear more constructive suggestions from our employees, and they are more likely to align their thoughts if our decisions are more transparent. If we make decisions behind closed doors, we will end up in trouble someday.

In late 2010, more than 60,000 Huawei shareholders from over 100 countries and regions gathered at over 300 polling stations and elected 51 representatives, who then voted to elect a new board of directors. The whole process was open and democratic, without any interference, and the event went rather smoothly.

For years, Huawei has convened brainstorming sessions, which are dubbed "Zhuge Liang Meetings,"[4] before any major decisions are carried out. Since 2010, such sessions have become even more frequent and they are conducted in a freer and more democratic way. Huawei is in the process of developing a decision-making model and a cultural climate marked by moderate authority and controlled democracy.

Huawei's relationship with the media has also greatly improved. In 2010, Ren Zhengfei said in an unusual tone, "Before the media, we used to act

---

[4] Zhuge Liang is one of the most respected strategists in China's history, a lifelong scholar, and a renowned statesman.

like an ostrich that buried its head in the sand. I believe that I myself can do that, but the company cannot. The company should be proactive, and time is right to start accepting more criticism."

From 2012 onwards, Huawei has essentially thrown open its doors and windows for all the world to see. The rotating CEOs and board members readily accept media interviews around the world; the company readily publishes new information of all types, expressing candid opinions on all sorts of "sensitive issues"; eventually, urged by international advisors, Huawei's "one and only ostrich," Ren Zhengfei also made his media debut in New Zealand in April 2013. He later began meeting with mainstream media reporters in France, the United Kingdom, and other countries to answer all kinds of questions relating to his personal background, political views, Huawei's share structure, corporate governance, succession plans, personal hobbies, and so on—covering nearly everything that could serve to demystify Huawei and himself. In June 2014, Ren also held a sincere and thorough communication session with the chief editors from over 20 Chinese media outlets.

On January 22, 2015, Ren accepted a live broadcast interview from BBC reporters at a closed session of the World Economic Forum in Davos, Switzerland. The US *Foreign Policy* magazine commented that the interview was one of the three major highlights of the forum.

From that point forward, Huawei has completely stripped away its so-called veil of mystery. Moreover, the board of directors further decided that Huawei should not be overprotective of its own information, because it did not have many secrets to keep. It was agreed that, whether it is general technology and market information, or most of the speeches and decisions made by the company's leadership team, whatever is publishable should be published on the company website. After all, in this era of Internet transparency, Huawei needs to learn how to play its hand right.

Compromise: The Law of the Jungle

Chapter 4

Lesson from History: Compromise Saved a Nation

The Historian's Visit to Huawei

On November 24, 2003, Qian Chengdan, a professor at Peking University and a distinguished scholar of British history, gave a lecture at the Political Bureau of the CPC Central Committee on the "History and Development of Major Countries Since the 15th Century." His lecture had caused quite the stir, not because the lecture was presented to the top leaders of China, but because he had revealed an unusual historical perspective.

One-and-a-half months later, Professor Qian visited Huawei and gave a lecture on essentially the same subject. His audience was over 800 mid-level managers and senior executives in the company, including Ren Zhengfei and Sun Yafang. Before he started, Professor Qian requested that none of the listeners record the lecture, take any notes, or disseminate his ideas.

About a decade has since passed and the ideological climate of China has been completely transformed. Professor Qian lectures in public much more often and his ideas have received mainstream recognition and approval. Under his guidance, in 2006, China Central Television (CCTV) produced a documentary titled *The Rise of the Great Nations*, which became a hot topic nationwide soon after it was aired. Ren Zhengfei bought 200 copies of the documentary and asked Huawei senior executives to watch it and discuss.

It was at the end of 2003, when Huawei had reached a critical juncture in its march as a new entrant in the international market. The lawsuit with Cisco was making this stranger's face more familiar throughout the global community. At this time, Huawei was facing a number of tough issues, cultural conflict foremost among them. Would the world accept this Chinese company, which was already building its reputation as a rule breaker? And how would Huawei participate in the international business arena, a game dominated by the West?

Professor Qian provided Huawei with a historical and cultural perspective that allowed the company to better envision its future. The leadership realized that by merely studying the West, visiting Western countries, and communicating with Westerners, they would fall far short of developing the international skills necessary to break into global markets, especially in Europe and America. Huawei had to acquire a systematic understanding of Western culture and institutions, and develop a way of thinking that could guide the company through the challenges it faced abroad.

How did Spain and Portugal rise to power? Both of them were small countries in the 15th and 16th centuries: Spain had only a land area of 500,000 sq. km and a population of 6 million; the land area of Portugal was not more than 90,000 sq. km and its population was only 2 million. But they had virtually occupied the whole world. Why? Their strength had come from their mercantilist agenda. They had navigated all the oceans and traveled far throughout the world. They would attempt to profit first through trade, and then plunder—and whenever business was not good, they would loot. They had, in short, taken the world through undisguised robbery and armed adventure, guided by a uniform national will and a strong primitive desire for expansion.

In the 17th century, an even smaller country took the place of Spain and Portugal. The Netherlands, which had only 45,000 sq. km of land and a population of one million, grew to become one of the major economic powers of the time. "The Netherlands had a nickname, The Coachman of the Sea," wrote Professor Qian, "which reveals a distinguishing feature of this small country back then: traversing the entire world by ocean, transporting goods to every place known to man, and making immeasurable profit. By 1700, the Dutch owned more than 10,000 merchant ships; their coast was lined with ports, and every port was filled to the brim with sea vessels, their masts stretching into the sky like an ocean-bound forest." They also created the earliest credit and financial systems and, most importantly, the earliest banks. In this sense, mercantilism was very mature in the Netherlands.

The Netherlands remained the center of the world until its decline in the 18th century, when it was replaced by the United Kingdom and France as world powers. According to Professor Qian, both the United Kingdom and France took two major steps when developing capitalism: moving from early mercantilism to late mercantilism and shifting from mercantilism to industrialism. Both were critical steps towards modernization.

The United Kingdom in particular had ensured its lasting power and prosperity through industrialization, which was a new form of capitalism. In the 17th century, the British population was only 4 million, about half that of Spain and a quarter that of France. But the United Kingdom became a global empire, the empire where the sun never set. At its prime, in spite of a still rather small population of 40 million, the British Empire controlled 50 colonies with a total population of 345 million and total land area of 11.6 million sq. km (96 times greater than the size of the United Kingdom itself).

Of the topics covered in Qian's lectures, what inspired Huawei's senior management the most was the Glorious Revolution. According to Professor Qian, "In 1688, England's Glorious Revolution avoided violence and war, and avoided mobilizing the people, while essentially resolving the issue of who would lead the country. And the revolution established the supremacy of parliament over the crown, setting Britain on the path towards a parliamentary democracy." This is the last revolution in British history, but there was no bloodshed or loss of life. What was the secret? The key was rational negotiation and compromise. All interested parties, including the king and the lords, had quarreled, bargained, threatened, and cajoled one another until they finally reached an agreement that would benefit everyone at all levels. They had substituted their tongues for guns, so to speak, and exchanged words for bloodshed.

Compromise is a word with negative connotations in Chinese, but it provided capitalist ideology with its most constructive philosophical underpinnings and ended up nurturing the strongest capitalist country in the world—the United States. Commenting on the congressional system of the United States, Hendrik Willem van Loon, an American author and journalist, said, "Compromise has saved a nation and founded an empire."

Fei Min, a former executive vice president at Huawei, started reading up on US history back in high school. When studying in Tsinghua University, although he majored in engineering, he spent a lot of time reading books and articles on the US system of government. He was particularly interested in the creation of the US Constitution: the stories, perspectives, details— and especially the political wisdom of the Framers of the Constitution—all deeply influenced his personality and the way he worked. When he had an idea, he would spend a lot of time trying to convince everyone who worked with and for him, and would consider a decision made only when he had reached a consensus among most of those involved. In his eyes, in its early stages Huawei was "like a miniature version of the United States"—people would heatedly bicker and quarrel behind closed doors, sometimes all the way till 2 AM or 3 AM. Once a consensus was reached, the decision was made, and when everyone left the room, it would be executed in with a united front.

In the eyes of the media, especially the Western media, Huawei was often depicted as a predatory, autocratic, and intolerant company. And it's clear that this very perception presented a serious barrier to Huawei's expansion in the international market. It is true that, for over a decade since its establishment, Huawei had acted aggressively like a hungry wolf or a pirate in order to survive in the cutthroat jungle. Huawei had, to some extent, been characterized by its aggressive and even suicidal approach to competition, and this aspect of its culture was restricting further progress.

In terms of Huawei's internal management, what Fei Min called the "miniature version of the United States" was objective and real. However, the fact remained that top management still called the shots and business leaders subjected the R&D and sales departments to their will. While this highly centralized style of management had helped drive efficient operations and rapid growth, it also led to many mistakes, especially when it came to staffing.

Huawei's chairwoman Sun Yafang said, "Huawei has been too rigid for years, and we are turning stiff. We need to make some changes."

So how did Professor Qian's story of the rise and fall of Western powers inspire a company from the East?

The Wisdom of the Jungle: Compromise Is Golden

Objectively speaking, from day one until the present, each stage of Huawei's development has been marked by compromise. Ren Zhengfei founded the company, but over time he has continued to dilute his interest in order to make Huawei an employee-owned company. Isn't that an act of compromise? Ren Zhengfei has compromised with tens of thousands of his colleagues, exchanging his interests for the unity, motivation, and stable progress of the company over the years. This is a rare case among Chinese entrepreneurs and even the global business community.

Some scholars compare this compromise of Ren Zhengfei's to a policy of buying people out. They argue that Ren is trying to lure everyone in the organization into the same boat by giving them material incentives. This is a boat that answers to a single authority, moving towards a single goal, with everyone fighting under one single banner. Who is this authority? Ren Zhengfei himself. What is the goal? To become a world-class telecom equipment provider. And the banner under which they fight? Customer centricity, dedication, and perseverance—Huawei's core values.

Ren Zhengfei has never disguised his intentions. He compares this practice to the division of spoils. He is the chief of a gang of pirates who would capture every boat on the sea and share the loot. Whoever seizes the most can enjoy the biggest share of the gains. As the commander of all pirates and a courageous fighter, the pirate chief is also in charge of distributing the bounty. Fairness is achieved only if the gains are shared on the basis of contribution. This would be especially attractive to the hungry and the needy, and followers would greatly increase in numbers. Indeed, in about 20 years, Huawei has built up a crew of more than 100,000 employees.

Excellent leaders must be willing to divvy up the spoils and embrace a spirit of compromise. But such a primitive culture of compromise and "sharing the wine and meat" is likely to give way to superstitious faith in a single person at the helm—an autocrat or dictator—causing the atmosphere of an organization to become oppressive and tense. Of course, an organization without any tension will most certainly collapse, but too much tension will also cause it to snap. In the time preceding and following the year 2000, Huawei was in a state of high tension, all authority centralized as it was in Ren Zhengfei. During this period, Ren was also extremely nervous and fearful. He recalled, "I was thinking about failure every day, and was blind to all success." He was anxious about the fate of the company and what would happen if it failed.

There were times when he would wake up in the middle of the night, wondering what he would do if the company was unable to cover its monthly salary payout of over CNY300 million.

Huawei, however, was capable of self-reflection and self-criticism. That is what sets the company apart from others. In 2000, aware of the company's dictatorial approach concerning the promotion and selection of managers, Ren Zhengfei proposed that compromise was necessary when considering management candidates. This was more of a requirement he set for himself than advice to other senior executives. He once said, "We can't always insist so strongly on our own opinions; we should listen to what other people think."

After Professor Qian's lecture at Huawei in 2004, the company began arranging for its middle and senior managers to learn about the modern history of Western powers through books, essays, and lectures on television. They learned that Spain and Portugal had fallen from prosperity because their culture of piracy deteriorated: with riches came hubris, and then extravagance, and then they were done for. And the Netherlands rose and fell soon after the country began to depend too much on trade and finance, which were not supported by the real economy: Overspeculation had destroyed the prosperity of the empire. The United

Kingdom rose to power and took their place, dominating the oceans and expanding across the world, launching an age of industrialism based on the manufacturing industry. More important was the Glorious Revolution, which substituted compromise for violence and served as the inception of modern-day capitalism. As Lord Acton of Cambridge University put it, "Democracy is gray, while compromise is gold. Compromise is the soul of politics, if not all." Generally speaking, politics starts with controversy and ends with compromise, and the Constitution of the United States is the very result of a compromise made over 200 years ago. Today, the American people are still awed by the rationality of the Framers of the Constitution.

Is compromise the soul of business as well? It is, at least partly so. "In the business community, there are no enemies; there are only counterparts and competitors," said Guo Ping, deputy chairman of Huawei, who is well versed in negotiating with Huawei's competitors in the West. Since 2011, Huawei has started calling its competitors "industry peers" in its public and internal documentation. After all, compromise is prevalent in all business deals, negotiations, partnerships, conflicts, disputes, and agreements, and it is the basis of shared success. And compromise is a necessary element in any internal decision-making process or cross-team collaboration, otherwise a company would be torn apart by infighting or autocratic rule.

In December 2007, Ren Zhengfei was in Hong Kong for a meeting with the former US Secretary of State Madeleine Albright, the minister for foreign affairs of the most powerful country in the world, known in international political circles as "tough," "having an iron fist," and possessing a "hawkish style." In the meeting, however, she seemed uncommonly sentimental: "Before we met, I read your articles 'My Father and Mother' and 'Huawei's Winter.' They left me with a strong impression. Human feelings are fundamentally the same." She also told a few anecdotes about her own father.

It was in that meeting, when asked by his guest, that Ren Zhengfei elaborated on the logic behind Huawei's growth and success. It was also the first time that Ren connected the three words "openness, compromise, and grayness" as the secret weapons of the company's rapid development from nothing to something, from small to large, and from weak to strong.

Half a year later, Ren Zhengfei elaborated on the correlation between openness, compromise, and grayness in a conference with company leadership:

> For some people, compromise may mean weakness and indecisiveness, and refusal to compromise is a heroic act. They believe that conquering or being conquered is the only relationship that exists between people, and there is no middle ground.
>
> 'Compromise' is a practical and adaptable law of the jungle. The wise ones in the jungle of humanity are ready to accept or seek compromise when necessary. One survives on reason rather than on impulse, after all. Compromise is an agreement reached under a certain circumstance. It is not always the best solution, perhaps, but it is the best option before the real best option appears. It has a lot of benefits.
>
> Compromise does not mean giving up principles or making unconditional concessions. A wise compromise is a fair exchange. To achieve our most important goals, we can compromise on minor goals. Such a compromise is not a matter of violating our bottom line, but making concessions to gain an advantage, and to achieve our goals by means of a fair exchange. Of course, any compromise is unwise if it lacks proper balance, misses the primary objective, or incurs unnecessary cost in reaching that goal. Wise compromise is an art, a virtue, and an essential skill of any manager.
>
> Only through compromise can we achieve a 'win-win' situation and 'multiple wins.' Without compromise, no party will win. As compromise helps eliminate conflict, refusal to compromise can be a prelude to resistance.

Several years ago, when we were drinking tea together, Ren Zhengfei said to me, "Should the east wind prevail over the west wind? Or should the west wind prevail over the east wind? Why can't winds from all directions sing together in harmony?"

War or Peace: A Matter of Pragmatism

Yitzhak Rabin or Ariel Sharon?

Business avoids extremes, and so does politics. In March 2001, Ren Zhengfei commented on Israel's prime ministerial election in his essay "Huawei's Winter:"

> What is a leader? What is a politician? In their last election, we witnessed short-sightedness among the Israeli people. Yitzhak Rabin realized that Israel was a small country, surrounded by millions of Arabs. Even though Israel has won many wars in the Middle East, it's possible that the Arabs will grow strong in the next 50 or 100 years. He believed that the Israelis would again be forced away from their land if they did not trade land for peace to achieve harmony with their Arabian neighbors. Should the Israelis lose their land, it's hard to say whether or not they would be able to regain it for another two thousand years. The majority of people only focus on their personal short-term interests. When leaders are adept at catering to short-term interests, they will win the support of the people. Thinking about this, I at once realized that we are no less myopic.

What is a good leader? A good leader is an outstanding person who is able to lead a country or organization out of crisis with independent thinking, great vision, and the ability to change with the times. Yitzhak Rabin was a good leader.

What is myopia?

When he met with Ren Zhengfei, British adventurer and businessman Simon Murray said, "The largest human migration in history took place after the Second World War. About 3 million Jewish people migrated from the former Soviet Union and Poland to the Middle East and the United States."

Jewish people are a miracle of organization in human history. For more than 2,000 years, they have been living in Diaspora, but they have held on to their common cultural roots, a strong sense of self-awareness, and a sense of crisis. Jewish people in every corner of the world share a common spiritual determination borne of collective belonging. This has been passed down for thousands of years as part of their cultural heritage. Perhaps it is because of this that the Jewish people play a pivotal role in our modern world.

Israel has been the spiritual homeland of the Jewish people for over 2,000 years, a homeland that Jews in Israel and in other parts of the world are determined to defend at any cost. This is a goal without compromise, for which Israel has been fighting with its Arab neighbors, including the Palestinians, for decades.

Yitzhak Rabin made a compromise: land for peace. As the descendant of a Zionist, Yitzhak Rabin would rather secure lasting peace for Israel than set his sights on a single town or piece of land. But he was assassinated and Ariel Sharon took his place, a man whom some have described as being rather shortsighted and hawkish and whom some have accused of butchery. After a series of tough policies were thwarted, however, Ariel Sharon chose to compromise in his final years in office, accepting the Roadmap for Peace and acknowledging that the millions of Palestinians "under occupation […] can't continue endlessly." In the end, although he did not possess the long-term vision of Yitzhak Rabin, Ariel Sharon essentially proved to be a pragmatist, an opportunist.

So should Huawei learn from Yitzhak Rabin or follow in the steps of Ariel Sharon? Ren Zhengfei said, "I am a student of Yitzhak Rabin." Although he only has 1.42 percent share of the company, Ren Zhengfei considers Huawei to be his life-long pursuit. In a certain way, Huawei is the testing ground for his management philosophy. From ancient to modern, both inside and outside China, he soaks up relevant concepts from the vast treasuries of thought and organizational practice throughout history, takes the ones that are most useful for Huawei, and adopts them into his overall philosophy.

From Yitzhak Rabin, one can learn long-term vision, and from Ariel Sharon, pragmatism. But no matter how visionary or pragmatic a politician or a business leader is, he or she needs to be able to compromise. Kneeling down before another can be a deplorable thing, but after kneeling down, you can always stand up again—and is that not also an act of greatness?

This is especially true in China. When Lenovo's Liu Chuanzhi gives a lecture with his shoulders held back as upright and stiff as a soldier, few people in his audience would be able to venture a guess at how often this godfather of a businessman—the same age as Ren Zhengfei—had to bow to all manner of twists and turns, even humiliations, throughout his long entrepreneurial journey.

In 2003, in the early stages of Cisco's lawsuit with Huawei, the strategy of Huawei's executives was to sue for peace—to lose a little was to win. At the same time, Ren repeatedly stressed that, in history, Han Xin himself crawled between another man's legs[1] in humiliation, but later grew to become a great general.

An Iron Fist and a Soft Hand: Attack and Compromise

When two armies meet, they have but two choices: fight or make peace. In the end, whether they choose to fight or make peace depends on the strategy and approach they take to reach their goals. Zero-sum games are the most common in sports, but in real military situations and in a business environment, fighting is an alternative diplomacy. Most only choose to fight when no chance of peace is in sight, and they hope that fighting will force the other party to accept peaceful terms. The purpose behind issuing a challenge or meeting a challenge, after all, is to gain the upper hand in the ensuing negotiations. Although in a school like West

---

[1] Han Xin is a famous general in Chinese history. The story Ren refers to here is a story from Han Xin's youth. Orphaned at a young age, one day he was heading down to the ocean to catch some fish when he was stopped by a young butcher. The butcher pointed to the sword that Han Xin was carrying and said, "If you've got it in you, run me through with that sword. And if you're too chicken, then get down to the ground and crawl through my legs." Han Xin, not wanting to be punished for killing the butcher, got on his hands and knees and did as he was told. This was an immensely humiliating act, but because he chose to preserve his life and freedom, he later grew to become a great general who helped found the Western Han Dynasty.

Point, their approach to military discipline and culture allows no room for compromise outside of the call of duty, we have to remember that that is for soldiers. On the battlefield, soldiers must go all out when they get the order to charge, and never stop until they secure final victory. Never surrender, never give up, don't look back, charge forward—this is single guiding principle for soldiers in the front lines. This is also true of the business world. At the front lines of the market, the sales force, from general down to solider, must be fully possessed of wolf-like aggressiveness, tenacity, resilience, and even a bit of an outlaw spirit, with no room for wavering or uncertainty.

In the command center, however, or among the decision makers, advisors, and directors in a business, they play another game. With a specific goal in mind, whether they be in the field of battle or business, they look for a way to bring an end to the deadlock: They balance and rebalance all the trade-offs until they reach a compromise that is mutually beneficial. Of course, such compromise is based on the uncompromising fight of the frontline soldiers. Because of this, compromise is the wisest way to achieve the greatest gains at the lowest possible cost. It should still be noted that all compromises, or even alliances, are temporary and changeable because they're typically aimed at one-off instances of carving up the spoils. After the spoils are divvied up, the players enter a new battlefield where they have to charge and fight again, and seek compromise yet again. It's possible that competitors change and newly arrived adversaries have completely different tactics, but the law of strategy remains: War is waged to create peace, and compromise is necessary for mutual benefit and lasting peace.

Wang Yukun, a famous management expert, once published a short essay about Huawei entitled "The Golden Mean: Decoding Huawei." It is well written, defying many conceptual labels that the media have applied to Ren Zhengfei and Huawei.

One of those labels is that Ren Zhengfei does not meet people halfway, nor does he allow his subordinates to betray principles in the name of compromise. Complaint and excessive moderation are deeply rooted negative aspects of Chinese culture, with far- and wide-reaching influence. Yet Ren Zhengfei does not complain, nor does he follow the crowd or take the middle ground. He is rational to a fault.

This label was given by Zhou Juncang, an author who worked as editor-in-chief for the internal publication *Huawei People*. As a journalist, Zhou has only seen a flat and superficial image of Ren Zhengfei; he is not able to see the multifaceted, round character of this business philosopher. In fact, Ren Zhengfei is a contradictory mixture of determination and flexibility, toughness and softness, sense and sensation, aggressiveness and compromise. Similarly, Huawei's culture is also a hybrid of East and West, tradition and modernity, conservativeness and innovation. And in truth, biologically speaking, crossbreeding tends to lead to a more hearty and vigorous species.

In an essay titled "From Philosophy to Practice," Ren Zhengfei said:

> China has long been influenced by the Confucian doctrine of the mean.[2] This doctrine has played a significant role in maintaining social order and stability, but has also deterred the growth of heroes and prevented them from driving society forward with their individual strengths.

And yet he has also said on several occasions that Huawei's success has been founded in avoiding extremes, and the traditional concept of the golden mean has kept the company from falling apart.

Liang Guoshi, a former employee, wrote a book about Huawei titled *Wolves Breaking Out*, in which he recorded a dialogue with Ren Zhengfei. In 1996, he and Ren Zhengfei took a walk under a snow-covered mountain in Bulgaria. Ren asked him, "Do you know why Huawei has become so successful?" He did not. Ren Zhengfei told him, "Because we have followed the doctrine of the mean."

Wang Yukun comments on this in his article:

> Many people see Ren as aggressive, stubbornly disruptive and obsessed with surpassing others. What they don't know is that these extremes are only on the surface. What's truly going on

---

[2] *The Doctrine of the Mean* is a Confucian text that deals with maintaining balance and harmony through moderation, sincerity, and general rectitude. This text, along with all Confucian doctrines, has had immense and long-lasting effects on Chinese culture and interpersonal relationships.

inside him is one act of regression after another, balancing and then rebalancing again. In life and business alike, strolling along a smooth, nicely paved and predetermined path simply won't suffice.

Theory is a gray thing, while life is like an evergreen. Theory is rich and changing, and adjusts to reflect changes in circumstance, sometimes an opposing force, sometimes unifying. You can't use a rigid or set approach to deal with the ebb and flow of evolving ideas. However, all changes must revolve around ensuring an evergreen life. Ren Zhengfei advocates compromise, but he also says, "We will not stand for compromise when it comes to dedication. Any employee who isn't dedicated, or who plays it safe out of self-interest—we need to be firm and let them go. Otherwise we have no way of ensuring long-term peace and stability." In other words, values must not be compromised. As the soul of the company, its values should be observed to the letter.

In another speech, Ren said:

> Openness, compromise, and grayness may not be as important for more junior employees. Of course, there's nothing wrong with their learning how to better manage interpersonal relationships, although I have often observed that they misapply these concepts. Senior managers, however, must absolutely learn these principles so we can achieve solidarity and unity from within.

What Ren Zhengfei has not explicitly stated is that, as a way of thinking, the doctrine of openness, compromise, and grayness is also a political culture within the organization. If it is well-developed and implemented, such a culture unites all senior executives together in an inclusive, open, cooperative, and mutually enhancing environment. At the same time, it effectively averts the territoriality, heated opposition, and cliquishness of royal-court politics.

Kevan Watts was the former co-head of global investment banking at Merrill Lynch and currently vice chairman of global banking at HSBC. He spent 35 years in the investment industry and worked with many investment banks. When comparing Huawei's management with that of US companies, he remarked that when a senior executive leaves an US company, the company cleans out a ton of people and reshuffles top management, which has led to a more tribal form of management. One person's ups and downs can affect the fate of many, and it's a painful process every time. In contrast, Huawei's quite different: Not only does it employ a novel rotating CEO system, but the senior executives who left the company still stick around to provide strategic advice.

Meanwhile, the doctrine of openness, compromise, and grayness is a strategic basis for business competition. Competitors are not necessarily enemies and competition doesn't have to be a zero-sum game. A business should engage its competitors to achieve mutual benefit.

Too much of an adaptive swing is no good for entry-level employees; however, senior leaders are strategists, so dialectical thinking is a must for them.

The choice between war and peace, attack and compromise, depends on a pragmatic balance of the company's short- and long-term goals. There are no hard and fast rules.

The "Bitter Vine" Strategy and Capitulation

Attack Is the Best Defense

The fight between Cisco and Huawei revealed the stress and anxiety that the rapid rise of Huawei had caused in the Western world. Around the year 2001, almost as if they were conspiring for a coordinated attack against Huawei, a large number of European and US companies staged a "patent siege," requesting 1–7 percent of Huawei's sales to pay for their patents.

Undoubtedly, this was an unprecedented challenge for a Chinese company that had grown up out in the boondocks. It was in this context that the idea of learning from Yitzhak Rabin, to "trade land for peace," came about. When a bull finds itself in a china shop, it either gets driven out or slaughtered. But isn't there a third option? As it turns out, there is: The bull needs to accept and obey the rules of the china shop and later become one of the rule makers himself.

Song Liuping, president of Huawei's legal department, once anxiously commented that a lack of fear had made Huawei successful, but being fearless, especially towards the law, would be a huge challenge as Huawei went global.

The "land for peace" compromise was led by Huawei's legal department, in which they actively engaged in negotiations with international companies to sign paid licensing agreements. Meanwhile, Huawei executives came to a collective realization that intellectual property is a "high-voltage line," a disruptive force with life-or-death implications for the company. Every layer of the company needed to have a proper sense of reverence for the law and, at the same time, the company had to put more effort into protecting its own intellectual property. After the lawsuit with Cisco, Huawei launched its "2008 Strategy"—from 2005 to 2008—to achieve a more balanced position with multinational companies in terms of IPR protection and management. A special leadership unit for IPR was set up at the corporate level, with a patent committee for each product line to select, examine, and decide on patent strategies. Some patents can be used as killer trump cards, powerful enough to destroy competitors and therefore potent as cold deterrents, while a large number of general patents can be used to strengthen and build out the company's patent-packed "wall of defense."

By the end of 2014, Huawei had applied for more than 70,000 patents, among which 38,825 were granted. Huawei is in the process of surpassing its Western rivals in terms of global IPR presence and overall capabilities.

Since 2010, more than 80 percent of Huawei's royalty payouts have been made to Qualcomm in the United States.

For any business, its ability to accumulate patents is fundamentally determined by the intensity of its investment in R&D. Scholars have insightfully pointed out that a company's financial statement speaks volumes about its sustainability and competitiveness in the future. Companies compete in terms of endurance, where marathon runners—not 100m sprinters—are the ones that win the race. Over the past

20 years, Huawei has consistently invested 10 percent of its sales revenue in R&D, totaling CNY200 billion between 2001 and 2013. In fact, after 2010, the company stepped up investment in basic science: its R&D investment accounted for 14 percent of sales in 2012, 12.8 percent in 2013, and 14.3 percent, or CNY40.8 billion, in 2014.

For many years, Huawei's annual salary expense, tax contribution, and R&D spend have far exceeded the net profit of the year, over time building the horsepower it needed to compete globally.

This horsepower is also reflected in its legal contests with Western companies. Since 2002, the company's legal department has dealt with around 2,000 cases every year, mostly related to intellectual property. Sometimes there were four or five litigations a day. Many were staged by "patent trolls" and quite a few involved patent conflicts with Western companies. Aside from softer skills like patent swapping, cross payments, and knowing how to step up to blackmail, the company has to play IP hardball from time to time, both inside and outside the courtroom.

One thing that might be of significant note to other Chinese companies is the fact that, over the past 10 years, Huawei has outdone Western companies in its ability to utilize legal resources in Europe, the United States, and China. Of course this is predicated on maintaining a favorable position in the main battlefield, the United States, where Huawei has learned and familiarized itself with how to leverage the routines and culture of a country's legal institutions. Out of some 100 lawsuits in the United States, Huawei has never lost a single case.

As Song Liuping observed, politics and media in the United States do not necessarily have to be fact-based, and neither do petitioning or lobbying. However, facts and evidence are definitely required in the legal system. Therefore, he has "confidence in the US judicial system," which uses a process of discovery to bring all facts to light. Anyone, based on a logical connection of the facts, can easily draw their own conclusions.

Facts are built on strength. Huawei is by no means a perfect company and there are instances where its employees do things that damage the company's reputation out of lack of fear for the law. However, this type of behavior is where Huawei draws the line, and whenever a case of violation is discovered, the company does not hesitate to expose the incident, subjecting it to judicial ruling and media criticism. Huawei's competitive drive in the global market must be built on transparency and strength.

During a meeting in Huawei's Shanghai Research Institute at the end of 2013, when a leader in the company's line of wireless products was boasting with great confidence about a defensive strategy to deal with technological evolution, Ren was disconcerted, replying, "To attack is the best defense." He then elaborated extensively on this point.

An aggressive advance is the only way to grow stronger every day.

Focus on Customers, Not Competitors

"Move the headquarters for 3G development to the Netherlands…" This was an order issued by Huawei's executives to its wireless product line in 2003. Never interested in short-term opportunities, Huawei had been investing heavily in WCDMA R&D for almost 10 years, placing the company in a passive position in the 3G market for quite a long time. As 3G licenses were not yet being issued in China, Huawei had to redirect its focus to Europe—Western Europe in particular. Europe had been a cash cow for Ericsson, Nokia, Siemens, Nortel and Alcatel: Both Ericsson and Nokia had a very high share of the market.

In response to that decision, Wang Tao, now president of Huawei's wireless product line, led a 15-man "crack team" to the Netherlands over 10 years ago. In January 2005, the company's executive vice president, Xu Wenwei, was appointed president of Huawei's European operations.

The Netherlands turned out to be a veritable beachhead for Huawei's wireless products in Europe. The project team was stationed in Room 808 of a local hotel, the "808 War Room" as it was called. Telfort, which was the smallest operator in the Netherlands, was acquired less than one year into the project. It was the first operator to have used Huawei's distributed base stations in the Dutch market, and after the acquisition, Huawei's base stations were replaced. However, the situation was later turned around when Vodafone Spain, in response to market pressure, began to explore strategic alternatives with a supplier from the East. By this time, Huawei had appeared on Vodafone's radar.

Opportunities are fought for, not waited out. At the time, Huawei was eyeing two projects at Vodafone: one in Egypt and the other in Spain. Some in the company thought that, because Vodafone had never worked with Huawei before, it was most likely to test them out in Egypt first. Not Xu Wenwei and Wang Tao, however. They had decided to "take Spain," a top-priority project that they believed vital—not only to win Vodafone's business in Europe, but also for the ripple effect it would cause globally.

The question was how to do it? They started out with system planning, as well as an extensive series of offensive and defensive exercises. They simulated every detail, from customer relations to technical support, from product service to project delivery, over and over again. They raised questions and meticulously thought them through: What would cause the customer to have no better choice than Huawei? Where was the strategic fit? In the end, the team came up with a slogan: "To Be Number One!" (Vodafone was number two in Spain, second to Telefónica.) The proposition was for Vodafone and Huawei to join forces, using Huawei's newly developed—and the industry's most advanced—SingleRAN solution to beat the competition and thereby secure a top position in the market.

The real problems had yet to come. At that time, Huawei had nothing in Spain. In fact, across Europe it had no reference project, no technical team, and no platform for service and delivery. How would the customer trust such a company? One of the most basic issues they had to deal with was the visas for experts flying in from China. As a preemptive move, Huawei went to the Spanish ambassador in China, requesting a letter to the client that the ambassador would give visa support if Huawei won the project.

Delivery was the biggest challenge. The first difficulty was shipping. Before 2008, most of Huawei's equipment was shipped to Europe by sea, or by air upon urgent request, from China, which would take 8–10 weeks, while its competitors, all based in Europe, could commit to delivery within four weeks. This forced the frontline team to "borrow" equipment from Huawei's Shenzhen headquarters, send it to a German distribution center two months before a potential contract might be signed, and then deliver it to the customers after everything was said and done.

"Concentrating a superior force to destroy the enemy forces one by one," hailed as the most important of Mao Zedong's 10 military principles, was employed by Huawei to its full extent. During the battle for Vodafone in Spain, veteran frontline commanders Xu Wenwei and Wang Tao pooled together the best account managers, product managers, technical staff, commercial negotiators, and delivery experts that the company had to offer. In the end, Huawei won a contract for 70 percent of the entire project, covering every major city in Spain.

"Surround, but don't attack."—This was a battle tactic created by Lin Biao, the late supreme commander and militarist of the People's Republic of China (PRC), during the War of Liberation in northeast China. Xu Wenwei's team also used it in the commercial battlefield in Europe: When an opportunity is not ripe, leave a small team out there to keep communication channels open, but only to surround, not attack. Xu further explained, "Meanwhile, I cooperate with your rivals and constantly appear in front of you. You will come to know Huawei through your competitors, and I will help them compete with you." When the time comes and all the right resources are in place, the whole army can swarm in and strike at once.

The strategy of "one point, two flanks" was also a battle tactic invented by Lin Biao. As early as 1992, Ren Zhengfei began advocating this battle tactic for market penetration. Fei Min and Li Jie had a profound understanding of this maneuver and, in the end, both were promoted exceptionally fast. So what is "one point"? It means cracking open a single point in the city wall when attempting to break into a fortress. You can employ any means to achieve this, be it grenade or missile, and try as many times as it takes. After a breach is made, the army should attack from both flanks at full force. So far, this principle has remained a classic for Huawei's frontline commanders around the world.

Huawei has a consistent style of attack, thanks to its deeply rooted ability to learn, including from sources like Chinese and foreign military management. Over the years, Huawei has learned a lot of valuable ideas, strategies, and tactics from the US army. Internal management jargon within the company includes two categories: the first is English acronyms learned from IBM, such as IPD and ISC; and the other is a massive military vocabulary, such as "battle," "battlefield," "front line," "rear," "rear service," "iron triangle," "navy seals," army general, gunfire, joint committee of regions (think about joint chiefs of staff), fighting to the top of the hill, and so on—a lexicon that has filled this massive business corps with the ability to boldly expand and execute its initiatives.

As a result, Huawei has won many "battles" all over the world, from Europe to Latin America, from the Commonwealth of Independent States to Africa, as well as most regions in the Asia-Pacific and North America. In 2013, Huawei's wireless products stood out among global competition, winning 142 new LTE (Long Term Evolution, or the fourth generation of mobile technology) contracts. At this time, all major European operators had established a close partnership with Huawei.

Huawei's sales outside China were only about US$500 million in 2005. When Xu Wenwei made his first appearance as Huawei's president of the European region, he set a goal for himself and his team: to achieve US$3 billion within five years. Three years later, in 2008, when he moved back to Huawei headquarters, his contract amount had already reached US$2.998 billion. By the end of 2013, that contract amount was again nearly doubled.

Huawei, an enterprise from the East, has made its presence in the European continent, where telecommunications was invented—not only as a challenger but, in a sense, representing a sort of external force, one that has shaken up the structure of traditional powers in Europe and has attacked their innate sense of superiority.

"Don't Be Lured in by the Red Cape"

The end of 2008 was when Ren Zhengfei first mentioned the "Bitter Vine" strategy. It was about six months after the global financial crisis had struck. Bitter vine is the common name for *Mikania micrantha*, a weed from South America, also known as "mile-a-minute vine" because its insane rate of growth far exceeds that of any other plant. Botanists consider it the terrorist of weeds, for it only needs a little bit of water and nutrients to thrive, at which point it suffocates the surrounding plants at lightning speed.

The Bitter Vine strategy marked the second time that a sharp maneuver by Huawei had enabled the company to overtake its competitors, who had slowed down to navigate a major paradigm shift in the industry. The first instance was in 2001, during the bursting of the IT bubble in the United States. It was with the same tenacity as the bitter vine that Huawei, a weed that had appeared out of nowhere from the East, burst forth at lightning speed out of asymmetric competition with Western companies to become a fearsome thicket in the telecom industry; and from there, in less than five years after 2008, it stepped up to a leading position worldwide.

However, Huawei's vine-like expansion along both product and market fronts has but a single orientation: the customer. Huawei focuses on customers, not competitors. For Huawei, benchmarking against competitors is not wrong, but it would be a huge waste of resources to make competitors the bull's-eye and deviate from the company's core value of customer centricity. In fact, those who prioritize trapping, defeating, or even killing off their competitors never really thrive, never really rise, nor are they able to reverse the course of fate when they enter into a falling trajectory. Putting the cart before the horse is never the true path to growth.

Needless to say, despite Huawei's focus on customers, the company's rapid growth certainly posed a threat to the traditional dominion and expansion of Western companies, which invited both friction and conflict. The Cisco–Huawei fight is perhaps the most representative of these.

For big companies, hurting each other is in their nature, yet reconciliation and moving on come equally as naturally. The reason is simple: No one can afford to lose. Huawei and Cisco had gone at each other, causing all manner of bumps and bruises, each demonstrating admirable wisdom in battle. Although the war had been started by Cisco—and Chambers is possessed of a uniquely American fighting spirit—the company's chief executive regarded Ren as an honorable opponent. In return, Huawei also treated Chambers with great respect.

One day in December 2005, about a year-and-a-half after Huawei and Cisco had reached a settlement, Chambers made a visit to Huawei. On his way to the bathroom, he had taken a detour through the employee cafeteria, where he mused to Guo Ping, who was accompanying him, "It seems as if you guys truly have what it takes to compete." And later, when he entered an R&D office, over 400 staff members stood up and applauded him, expressing their respect for this fearsome competitor. He hadn't realized that even Huawei's entry-level employees would harbor so much respect for him. He understood then that Huawei's leadership was rational and was moved by the fact that, without any preparation whatsoever, he was warmly welcomed by all strata of employees at Huawei.

Compete, but don't ruin the relationship—that has been Huawei's approach. The most recent meeting between Ren and Chambers was at the Beijing Capital International Airport in 2014. Towards the end of the conversation, Chambers invited Ren to visit the United States.

Ren chuckled: "I have no plan to visit the US just yet. The US is a small market."

Chambers replied, "If it's a small market to you, then I'll rest better at night."

He then invited Ren to be a guest at his house. Ren smiled, then asked for a rain check: "Let's have coffee in Europe first. Europe is much more romantic."

Contrary to the protectionist policy that the United States had adopted towards Huawei, its more romantic counterpart embraced the company with open arms. Respect for fairness and strength is part of the European tradition. Political dignitaries in the United Kingdom, France, Germany, and the European Union have met with Ren Zhengfei and other Huawei executives on multiple occasions. Huawei has also invested heavily in many European counties, building research and development institutions; meanwhile, it has developed joint-innovation centers with major operators throughout the Continent, winning a large number of contracts. Europe had become Huawei's own cash cow, a market of strategic significance.

And the question remained: how to alleviate the concerns of its European rivals? As early as 2009, the second year after the company had adopted the Bitter Vine strategy, Ren warned internal staff quite vocally that Huawei should not try to "dominate" the world. Instead, they must advocate compromise. Once, France Télécom asked Huawei and Ericsson to replace Alcatel-Lucent's equipment, and Sun Yafang flared up at Huawei's head in Europe: "Have you gone mad? Swap out Alcatel-Lucent's equipment in France? Do you have any concept of a business ecosystem?" Ren Zhengfei also called him, prohibiting the project. The ready-to-sign contract was abruptly called off.

In 2013, Huawei had a country office sign an agreement with a local operator to swap out Ericsson's base stations for Huawei's newer generation of equipment. When the board of directors learned about it, the country office and the head of the operator business group were harshly criticized. The board instructed them to write a letter of self-criticism for internal circulation and then informed senior management at Ericsson.

An Ericsson executive once said that if the lighthouse at Ericsson went dark, Huawei and the entire telecom industry would lose their bearings for the future. It was Ren's view that Huawei had to make it clear, both internally and externally, that Huawei doesn't aim to topple any other company's lighthouse; instead, Huawei is set on building its own lighthouse, while working to ensure that the guiding lights of Ericsson and Nokia would never go dim. Huawei would not dominate the world. As a matter of fact, Huawei was the one that put forward the idea of "industry peers" in the first place.

At the venue of the 2012 annual market conference, a banner caught people's attention: "The success of European companies in China is in line with Huawei's strategic interest." In recent years, Huawei has been cooperating with authorities in China's government to promote greater access for European telecom companies in the Chinese market. Their efforts have paid off. The 863 Program of the Ministry of Science and Technology (MOST) and major technology projects of the Ministry of Industry and Information Technology (MIIT) were opened up to European companies for the very first time. Companies like Ericsson and Nokia Siemens Networks were allowed to participate in the 5G (fifth-generation mobile communications technology) project funded by MOST; Nokia Siemens Networks participated in a major wireless project for the first time; Shanghai Bell and Alcatel-Lucent were given access to the same wireless project and part of the LTE R&D project.

Compromise begets more space for survival. In the several anti-dumping and anti-subsidy investigations that the EU Trade Commissioner had launched around 2012 against Huawei and ZTE, European companies like Ericsson and Nokia stood up to endorse Huawei: Huawei did not dump products in Europe. Nevertheless, Ren Zhengfei stated on many occasions, including in meetings with Chinese government officials, "Huawei wants to raise a white flag—in most cases, I surrender! We want to raise our prices to be as high as Ericsson's. Europe is a well-stocked granary for us. We can't use the hard-earned money from our brothers in Africa to subsidize Europe. If I earn more in Europe, the pressure on our brothers in Africa will be less, and we can help balance out the European business environment."

In February 2014, at the GSMA Mobile World Congress in Barcelona, Spain, at Huawei's magnificent gala reception for its customers, a British gentleman was talking to me about Spanish bullfights. At the end of the conversation, he said meaningfully, "Don't be lured in by the red cape—that's the sure path to getting killed."

That day, we also discussed the crisis in Ukraine.

It's Not the Story of One Man's Fight

Dictatorship and Beneficial Autocracy

Being a leader entails making critical decisions. A lot of leaders display ostensibly similar characteristics: decisiveness, even arbitrariness. And they are often dictators in their own right. Steve Jobs once said that collective decision making is stupid as it delays time to market and is a way to shirk responsibility. Steve Jobs had a classic dictator's personality.

In the true sense of the word, Adolf Hitler is globally recognized as a dictator. When several hundred thousand German soldiers were fighting against the Allies, Adolf Hitler was chewing the carpet in the basement of a castle, filled with great anxiety and dread. As a critic once noted, "The souls of leaders are drenched in stress." But when he appeared in public, Adolf Hitler always appeared determined, resolute, and unequivocal; He did not allow any room for doubt.

The founder of Hughes Aircraft Company, Howard Hughes, had led a dramatic and sensational life full of adventure. Named "Hero of the Century" by the American people, he had a despotic and compulsive personality. At times, he would fly his aircraft around the globe, narrowly escaping death on several occasions. At other times, he would shut himself up in room and refuse to bathe for weeks, using tissues to wipe his body clean.

Business organizations resemble a contingent of troops. There are a lot of similarities: Both require their members to fulfill appointed goals and tasks within an appointed window of time and in a given environment. So, it can be said that business leaders naturally share a certain idiosyncrasy with army commanders: They are dictatorial and authoritative. In other words, only those who possess this disposition can assume the role of a leader. It's no surprise that a great number of outstanding US business leaders once served in the armed forces.

Huawei's CISO John Suffolk, is from the United Kingdom. When talking about Ren Zhengfei, he mentioned that in both the United States and the United Kingdom, experience in the military is something to be proud of. Huawei should also be proud that its leader was once a soldier.

Both business organizations and the military need "beneficial autocratic rule," which is rooted in people's natural inclination for conformity and obedience. Conformity is a common phenomenon in the biological world, not just in human society. Wild geese fly in a V formation because a lonely goose would feel insecure; sheep live in flocks; wolves would rather hunt in packs even though they are extremely independent; and ants follow a strict order partly because they are reliant on the other ants and the colony. Organizations have appeared because most individuals lack a sense of security, direction, and confidence if left to their own devices.

Organizations are for the natural product of a culture of conformity. However, to be a conformist does not mean that one follows the majority; instead, conformity is following a single individual who leads the majority or the organization. The leader is like the queen bee, the lead wolf, the head sheep, or the ant commander. The leader must be able to keep a clear mind and steer the pack in the right direction, create hope for followers in dire straits, and lead them through to the other side. In many cases, the leader must make decisions alone, unable to depend on anyone else in the organization, as everyone else might lose composure during times of crisis. Moreover, at such critical moments, too many opinions would be a distraction and the best opportunity may be missed if the leader is not assertive enough. Making one or two such mistakes is forgivable, but making continual and repeated mistakes would certainly diminish the leader's authority and reputation, and fewer people would be willing to follow.

In this sense, leaders have become dictators because of the collective noncommittal nature of the majority of people within the organization. Whoever dares to make decisions and assume responsibilities has the potential to lead. This is very apparent in Western corporations. The West boasts political democracy, but their companies implement a totally different culture—"CEO culture"—in which CEOs have supreme power in strategic planning, personnel appointment, and operational decision making. Therefore, the fate of a Western company often depends on a single person.

Howard Hughes achieved great success thanks to his willingness to risk everything on a single decision, but he missed out on a lot of opportunities for the same reason. Steve Jobs founded the Apple empire with his stubborn autocracy, but it was precisely because of his excessively dictatorial approach that he was kicked out of the company the first time around.

There was a time when, in the eyes of journalists, Ren Zhengfei was a mysterious dictator. He was at once a tyrannical leader and an ill-bred boor. He might lash out at the company's leaders in public, putting them on the spot with coarse language, causing them embarrassment. He had even lost his temper when meeting with government officials. At other times, he would go so far as to kick out at colleagues around him. During interactions with others, he had little patience and would dominate the discussion, turning the conversation into a monologue and continuously interrupting others.

The fact remains that Ren Zhengfei is a charismatic and engaging leader. He enjoys exceptional authority and influence among more than 100,000 employees at Huawei. He is the highest authority in the decision-making panel. This is perhaps why he has been assertive and tyrannical, especially during the first 10 years of the company. While he can be rather liberal or even hands-off in decisions about the business and people, he at times becomes overly adamant in his opinion on appointing or using certain leaders. In some cases, his insistence has caused significant setbacks for the company.

Putting Power in a Cage

The world is changeable, although it operates according to rules. Any individual and organization can be assertive or seek compromises. A good leader should be adept at offense, but he or she should also be able to compromise when necessary. Compromise may not mean exactly the same as passive defense. Compromise is a practical art of communication and a form of democracy, which in some sense means to find the right resonance between louder and weaker voices.

At first, everyone is speaking their mind aggressively and loudly: They are trying to get their ideas and emotions across to everyone else, irrespective of what others say. Then, they begin to draw comparisons: who speaks louder, whose pitch is higher, whose opinions are more reasonable, and who has a bigger say in the group. In the end, compromise is the core theme of the democratic orchestra. Showy musicians in the band tend to produce cacophony, while those who compromise to facilitate a resounding performance are worthy of respect.

The most important member of all, however, is the conductor—in a business organization, the top leader. The leader can neither rush in conducting the performance nor give a solo performance on the stage. The leader must keep the bigger picture in mind and then pool the wisdom and strength of other members around it. To achieve the goal of the organization, the leader has to be the first to seek compromise when necessary and the first to command the team to charge forward. An organization may be powerful only if every member acts in unison under its leadership and compromises in the spirit of rationality. Therefore, the best decision is not often the most appropriate; being reasonable is usually the best choice. This is the efficacy of and the price paid for compromise—the necessary tradeoff.

George W. Bush, the former US President, believed the most precious invention of the human race is to tame power and put it in a cage. As the top leader of the most powerful nation in the world, the US President seemingly possesses supreme power. However, in the US political system, any leader is entitled to only a certain amount of power. During the eight years that Bush was the "master" of the White House, he was not able to unilaterally make decisions on domestic or foreign affairs. His decisions were subject to the approval of the Senate and the House of Representatives, which are always involved in endless bipartisan debates and disputes. In spite of his cowboy impulse, George W. Bush had not been able to break free from the shackles of the cage. His hands and feet were tied on many occasions as he wrestled with many actors; he had to make decisions and take actions within the framework of compromise.

Theoretically, politics differs from business. Politics concerns the masses while business has a CEO culture. In practice, however, a business is also subject to a governance system comprised of a board of directors and a board of supervisors. Corporate dictatorship is a myth; so is the idea that Steve Jobs never compromised. Some have even brutally commented that Steve Jobs died at the right time, the argument being that if Jobs had survived for another decade and continued as Apple's CEO, his arbitrary sense of heroism and his closed approach to product development would have ruined the company. Perhaps, and only perhaps, the success or failure of the company would have rested with Jobs. However, this is not a rare case in the business world.

Ren Zhengfei has admitted that Huawei used to operate under a highly centralized hierarchy. That implies a lot of risks, and the biggest risk is that the company dies if the leader goes. He said, "We can't bet everything on the survival of a single man. It is fortunate that I had not jumped from the building a few years back; otherwise, I wonder what would have become of the company." Therefore, in the coming years, as he argued, "Huawei must rationalize its management through culture, systems, and processes. The company should decentralize its decision making and strengthen its system of supervision. The two wheels must go forward in parallel."

Everything depends on the awareness of the top leader, of course. Leaders must get rid of imperialistic or totalitarian mentalities so that their companies can develop a culture of checks and balances. Ren Zhengfei said:

> Personal will has to be filtered through collective decision making. A leader must learn to make decisions in an open environment where he is subject to restrictions. He must not be afraid of losing his face, as I believe his face is not the topmost consideration.

In that vein, when he meets with disagreements at executive management-team meetings, he insists strongly on his ideas, but reaches a consensus in the end. Ren once said, "I don't expect all my requirements to be satisfied. I am often challenged."

In 2004, a principle was set by the top management that any executive appointment should be subject to a collective decision of unanimous approval. After 2009, a system of "separation of powers" was further established. The AT (administrative team) of the staffing department has nomination and veto-to-nomination powers; the AT of the supervisory department has evaluation and review power; and the ethics and compliance committee has the power to veto and impeach. If all goes well, the appointment will then have to be publicized on the company intranet for 15 days before becoming official. On top of this, there is a two-year grace period for backtracking. If the appointed person proves to be severely unsuitable, the AT of the staffing department will have to take collective responsibility for the consequences. This means that no single person in the company, not even Ren Zhengfei, has the final say about employing any executive; every member of the decision-making panel must compromise with the entire group.

In fact, a few leaders that Ren favored were vetoed for promotion by majority rule via this collective decision-making mechanism.

On January 15, 2011, Huawei convened a shareholders' meeting, during which its new directors and supervisors were elected. After the election,

Ren Zhengfei gave a speech that described the concept behind a newly established symbolic leadership role for himself and the chairwoman:

> In order to help our executives grow up better and faster, our chairwoman and I will sit on the board as symbolic leaders. Our role is to veto any proposal that we don't agree with and force the board to come up with a better alternative. If we believe the new proposal is still not acceptable, we will reject it again and ask them to involve more people in the discussion. In this way, we may not be able to make quick decisions like we did in a small team, but we will perhaps be able to avoid some major mistakes. This will enable everyone to take the initiative and proactively go out and manage; our veto power will serve to help everyone. [...] In the future, the chairwoman will be mainly responsible for impeaching managers, whereas manager selection will still be completed according to a system and process. In the development of Huawei's corporate system and culture, I will play a more supplementary role. Of course the symbolic leaders must be able to endure loneliness, otherwise the management team will certainly suffer. Any decision involves a long process of percolation, maturation, decision making, and execution. I must refrain from interfering with this process, because I will disrupt it if I keep giving instructions. Loneliness is painful, but it will be worth it if it can help Huawei succeed.

As the top leader of Huawei, Ren Zhengfei has consciously put his power in the cage, and as a result, Huawei's management has become more democratic and rational. Ren Zhengfei said:

> We are in a great time. None of our major competitors can implement the flexible mechanisms that we have at Huawei, while our smaller competitors do not possess the brand power that we do. We should be able to seize any opportunity as soon as it appears.

Cisco's Chambers cares quite a bit about when Ren Zhengfei is going to retire, and many others, including journalists, have bound the success or failure of Huawei with Ren Zhengfei in their minds. This is clearly an erroneous assumption. Huawei's future does not depend on a single man, but on a group of over 100,000 people who are fighting together as a team.

The Philosophy of Grayness: Fueling and Controlling Desire

Chapter 5

The Diversity of Human Nature

"Seeking Truth from Facts May Sound Easy, but It's Really Hard to Do"

Over 20 years ago, I was a psychology lecturer at a normal university. During one of my lectures, I gave a magnifying glass to some students, asking them to observe their skin and describe their observations. Soon a girl cried out, "I can't stand looking at this!" The other students had the same response: It gave them the creeps.

What followed was a discussion on perfection and how no one is perfect: Every person, including great leaders, is multidimensional. They are kind in one aspect and cruel in another; beautiful in one aspect and ugly in another. A good person has imperfections, whereas so-called bad people might have their commendable traits too. Whether an individual is good or bad depends to a large extent on the perspective from which they are observed. Therefore, I concluded that first, you cannot take a clear-cut position on anyone: You should not love or hate too easily, or you will lose friends and peace of mind. Secondly, since no one out there is a full-on saint or a full-on demon, you don't have to look up to anyone but you can't look down on someone either. Thirdly, you must treat yourself properly—vanity and self-deprecation are extremes, neither of which are beneficial for personal growth.

This idea was ill-regarded in the 1980s. Even today, without proper context, the argument that there is no crystal-clear distinction between right and wrong is still subject to criticism. But as I see it, many Chinese people put themselves at odds, harbor hostility towards one another, and suffer tension and anxiety not just because of unequal distribution of wealth, but mostly because our culture lacks tolerance. We don't accept differences, including those concerning background, personal disposition, social status, and beliefs. We simply aren't used to looking at issues from another person's perspective, and as a result, our perceptions are stereotyped: People fall into definite categories, not unlike the different roles in Chinese opera.

Of course, art is different from reality. The reality is that people are diverse and they lead diverse lives. Whether you're a nursery teacher taking care of dozens of kids or the head of a state governing millions or even billions of people, you face a similar problem: No two people look or act alike. The world is varied and complex. In his 2012 New Year address, Dmitry Medvedev, then the President of Russia, said, "Yes, we are all different, but this is precisely where our strength lies." A villager in a remote area of China reflected after running for a seat in the village's governing body, "People's hearts go deeper than wells. If I can manage this village, I can also manage a county."

Human nature is like the universe—vast but tiny, constant but changeable, regular but diversified. It is constant because human evolution is based on hereditary genes, but every man is rendered unique by his educational background, home environment, and personal experience. Therefore, the most challenging task in the world is to recognize, select, transform, and employ people, and combine them in the right way to achieve a particular goal. From empires to regions to families, all forms of organization must have a solid grasp on our slippery human nature.

By the end of 2014, Huawei had more than 170,000 employees (including nearly 40,000 foreign employees), of which 90 percent were college graduates, with an average age below 30. Unlike other Chinese companies, Huawei has a unique mix of employees: most are young intellectuals and they come from different countries. Therefore, Huawei faces a pressing imperative to understand and manage human nature.

When you go deep into Huawei's operational model, management system, organizational structure, corporate culture, and the history of its transformation, you will surely find that Ren Zhengfei is a master of human nature and that Huawei's success does not come from technology, from the market, and most certainly not from resources, but from an insight into human nature and the ability to harness it.

Although human nature is complex and mercurial, and any organization that consists of many individuals is, by nature, even more complex and mercurial, Ren Zhengfei has bundled everything together in a concept he calls "grayness": If you can't penetrate its depths, you'd better recognize it, tolerate it, and even grow to appreciate it.

The poet Johann Wolfgang von Goethe once said, "Viewed from the summit of reason, all life looks like a malignant disease and the world like a madhouse." But what if we were to look at it from a different perspective?

Ren Zhengfei said:

> If the managers we're selecting through our competency review process are all extremely flawless people, then they're clearly saints, monks, or even priests. This isn't what we're looking for. What we're looking for are strong fighters who can form an army. Our competency review process is an objective process of evaluation, a huge improvement on management in the past, which was too emotionally driven. But then again, an emotional management system has its merits: it doesn't demand perfection, or rather, it doesn't require everyone to be a perfect person.

Early in 2015, Ren Zhengfei stated the following in a meeting with the executive leadership unit under the human resources committee:

> Our leadership policy needs to be designed and implemented with more shades of gray. Don't shun imperfection or look at things as either black or white. It's only natural for managers to be flawed; no one is perfect. If it weren't for our insistence on a more lenient policy, we wouldn't have 4,000–5,000 people voluntarily reporting accounting violations. This shows that our lenience in policy has worked, and we need to preserve this moving forward. We are not enemies with one another; we are

brothers-in-arms, together in the trenches. If someone gets soiled fighting in the trench, it's no big deal—just wash it off. Don't try to be so spotless, at least not within this unit. If we always look at people through shaded lenses, then good people will cease to exist. The idea of "openness, compromise, and grayness" is something you all need to reflect on, over and over again.

Ren Zhengfei asked the executive leadership unit to become a bridge that connects the company with its managers, a mechanism that "hears their voice and eases their pressure" by providing the company with negative feedback on the managers' behalf. Ren wanted the unit to "learn to side with the managers rather than the company," to become confidants, not spies. "As for some of the things you learn from them," he said, "just let it rot in your stomach—there's no reason to let it out."

At the same time, Huawei has a clear set of philosophies for its auditing and supervisory systems: The purpose of supervision is to support business operations and commercial success by deploying the company's troops at full force. A supervisory system has to be a system of compromise, a conciliatory system, one that acts as a cold deterrent. Yet its ultimate goal is to allow the business to grow.

Ren Zhengfei requires his auditing staff to adopt an accommodating mentality in order to "seek truth from facts." As Ren puts it, "Seeking truth from facts may sound easy, but it's really hard to do," and he went on to encourage his auditors: "Put yourselves in the shoes of the people being audited, rather than just doing a job. If you're auditing for auditing's sake, we won't meet our objectives."

For Huawei's top management, the goal is to unburden its people, get the machine running, and unleash its troops at full force to "take over the hilltop." And what does the "hilltop" signify? To reach US$100 billion in annual sales within five years? Or to explore uncharted territory in the global ICT industry? Whatever the hilltop is, in order to get there, every one of "the home-bound heroes in the evening mist"[1]—Huawei's 170,000 employees—has to perform at their best.

---

[1] Here, the author quotes the poem "Shaoshan Revisited" by Mao Zedong, in which Mao visits his hometown after an absence of 32 years and is pleased at the sight of so many heroes, or farmers, heading home after a day's work in the bean paddies that stretch as far as the eye can see.

Ren Zhengfei once used the *bagua*[2] diagram in Chinese traditional philosophy to explain his grayness theory. In the center of the trigrams, there is a circle known as the Supreme Ultimate that consists of a black fish and a white fish, also called the yin fish and yang fish, respectively. The white fish has a black eye and the black fish has a white eye. This means that yin contains yang and yang contains yin: They both reinforce and neutralize one another. When applied to an individual, this also means that the person has strengths and weaknesses that both offset and complement each other. The glory and destitution of an organization are rooted in the same concept: at once causing the other and existing because of the other.

From the perspective of Western history, Ren Zhengfei elaborated further:

> Protestant ethics reveal the philosophical spirit of religious reform in the Middle Ages, which liberated humanity and later evolved into the spirit of capitalism. It's a liberal spirit that embraces and protects personal differences and human rights, which unleashes and motivates human potential, driving progress. […] Capitalism has two major assumptions: the first is that man is selfish; and the second is that man is greedy. This system restrains human defects and sets our motivations free.

Building on a foundation of deep insight into human nature, Huawei's leadership team has designed and disseminated the core values of the company: customer centricity, dedication, and perseverance.

With these core values, they have simplified and clarified the diversity, complexity, and volatility of human nature. Every person in Huawei, no matter how many shortcomings or personal defects they have, can find

---

[2] In traditional *taiji* cosmology, the *bagua* (which literally means "eight divinatory symbols") is a visual arrangement of eight concepts or forces that represent the interrelated fundamentals of nature and reality. Each force is depicted with a trigram: a set of three broken or unbroken lines that, together, represent the essence of the force. The interrelations in *bagua* are applied across a number of disciplines, including Feng Shui, traditional Chinese medicine, divination, and astronomy.

their own niche as long as they are given the right combination of material and spiritual motivation and as long as they recognize the principle that those who contribute more, get more.

This "gray approach" to management was pioneered by Ren Zhengfei, who was the first to put it into practice.

Intellectuals, Soldiers, Heroes, and Generals

In 1990, 23-year-old college graduate Hu Houkun took his first job with Huawei as a maintenance engineer. Ren required all members of the sales team to start from maintenance engineering, which was a psychological challenge for young graduates, filled as they were with the superiority born of knowledge.

In 2001, Huawei's top management made a wrong judgment on China's 3G systems. They believed that "the country would soon issue 3G licenses" and that 3G was a great opportunity for the company to overtake its competitors. It hired 6,000 college graduates that year alone in order to realize CNY10 billion in 3G product sales the following year. By early 2002, however, they realized that they had sped right into an unfavorable situation—the 3G licenses weren't going to be released in the foreseeable future. So what were they supposed to do with the 6,000 new hires? It was finally decided that they would be transferred into the customer-service system to perform various maintenance roles. Many new hires, however, were not happy about this, arguing that they were hired to do R&D, not "hard labor" like customer service. In a state of intense agitation, someone ran to the podium and snapped at Ren when he was hosting a dialogue with the new hires: "Were you insane to hire so many people in one fell swoop?"

For an individual, going from a "man of letters" to a "man of battle" is an important step in becoming part of the Huawei organization. For Huawei, this is a process of talent screening. It's like sending the crashing waves of the Ganges through a giant sieve to filter out hundreds of millions of grains of sand, gradually panning out the thick-skinned, tough-nerved, and strong-minded pieces of diamond and gold. Batches after batches of men of letters, including those born in the 1980s and 1990s, went through this process and later became the soldiers and generals of Huawei.

Jiang Xisheng, secretary of the Huawei's board of directors, visited a county-level postmaster in China 20 years ago. He was too timid to even look at the postmaster when he entered the door. Hurriedly placing company literature on the postmaster's desk, he blurted out, "Mr. Postmaster, these are our company's materials." Blushing, he then turned and ducked out the door. About a year later, he turned out to be the first person in charge of sales at Huawei. Hu Houkun and Mao Shengjiang were both his protégés.

Ren Zhengfei's saying that "being shameless is the way to progress" comes from that period in time.

Hu Houkun once waited in the corridors of the Baise county post office in Guangxi province for two days just for a chance to get in the door and visit a customer. Li Jian (currently the vice president of Huawei's Joint Committee of Regions) used to be a product manager in Nigeria. He painstakingly sought an opportunity to meet up with the president of a potential customer's organization and was finally allowed to see him. However, he had been kept waiting for more than three hours outside the president's office and eventually cornered him outside the bathroom door in order to secure a meeting.

Li Jian also joined Huawei right after he finished his master's degree. In school, he was the student-council president as well as a committee member of the provincial student federation. However, once in Huawei, "everyone has to start from scratch because performance is the only means of evaluation." Any instances of past glory would be "reset to zero," so everyone had to go through a self-deprecatory process of conquering vanity, arrogance, cowardliness, boastfulness, and laziness.

In Nigeria's sweltering 40°C weather, Li Jian was always suited up, carrying a laptop and a projector, visiting clients from dawn to dusk. Within three months, he signed contracts in excess of US$30 million; a year after, his contracts amounted to nearly US$200 million; and in the third year, over US$400 million. For four consecutive years, sales in Nigeria ranked first in the world at Huawei, so the Nigerian market was nicknamed "Shangganling Hilltop"[3] within the company. Li Jian was promoted three times in a short period of time, which was uncommon, rocketing from a low-level product manager to the vice president and then president of the West Africa region, managing over 2,000 people.

"Do you know how to giggle?" Peng Zhongyang, Huawei's former president of the North Africa region, coached an introverted employee who couldn't speak English very well. "If you don't know how to speak, or if you're too afraid to speak, then just smile at customers and giggle a bit!" Laughter is a surefire way to narrow the distance between people. The new hire, thus coached, took his advice and put on a smile every time he met with customers. As a result, he soon gained their trust and was able to land a few contracts.

Wei Chengmin, now president of the South Pacific region, was staying in Henan province in 1996 as an after-sales engineer. Once, along with a colleague, he traveled across the snowy countryside for three days and nights by bus, on a cart mounted on a walking tractor, and then in a rental car. They spent their nights in the mines and villages, maintaining switches for clients. They ate in the villagers' homes when hungry and napped in the machine room when they needed rest. One time they even spent the entire month living on the client's machine-room floor.

And thus skins were thickened and willpower was strengthened, while people became tougher and more adept at roughing it like outlaws. In Huawei's early days there were eight "bandit kings" who shared the same traits: Gregarious and omnivorous, they drank deeply and ate merrily; they would pound on their chests with braggadocio and fight hard in battle; they were not afraid to go against authority; they loved their soldiers but were hard on them… and before they joined Huawei, most of them were outstanding students from well-known universities in China or came from key roles at state-owned enterprises. Most of them were more or less men of letters. But in Huawei, they were transformed into

---

[3] The Shangganling Campaign is better known outside of China as the Battle of Triangle Hill, a fierce month-and-a-half-long campaign during the Korean War. Although viewed differently by those on the other side of the battle, in China it is often celebrated as a successful act of endurance and courage against all but impossible odds.

men of battle—with a dash of the hero or the daredevil thrown into the mix.

Over the past 20 years, Huawei's ranks of "generals and commanders" have been lined with hundreds of these daredevil types.

That being the case, as their leader, how did Ren Zhengfei fuel their ambition, channeling each individual's willpower into a collective, cohesive force that would be competent in the battlefield?

Sit around and Dream Big, Rise up and Fight Hard

So what motivates people at Huawei?

Is it the employee shareholding scheme? Is it money?

Undeniably, in any commercial organization where maximizing profit is the goal, the pursuit of financial freedom is the fundamental driving force for employees. Huawei's 27 years of rapid expansion had an important prerequisite—the financial-incentive and wealth-distribution system that centers on dedicated employees, or what the company calls "fighters."

However, if it was only about wealth distribution, at most Huawei would gather a motley crew. An organization has no soul if its people are in it only for the money; and a soulless organization is not sustainable.

One of the core motivational forces at Huawei is to talk big—to sit around and indulge in tall talk.

Zheng Baoyong used to be Huawei's "number-two man." When asked about his first impression of Ren Zhengfei, he answered: "This guy's full of it!" I turned the question around: "And what about yourself?" Zheng chuckled: "I'm even more full of it."

Li Jie recalled a strong memory:

> We learned to talk real big at Huawei. In the early years, our boss and Zheng Baoyong were bragging in one room while a bunch of us were blowing hot air in the other. We were so full of crap, but we really pumped each other up.

What were they going on about? At a time when Huawei's annual sales revenue was less than CNY100 million, Zheng Baoyong boasted that CNY100 billion was right around the corner. Ren Zhengfei would roll up his pants legs, roll up his sleeves, and paint the following picture for that group of fresh college graduates: "In 20 years, Huawei is going to be a world-class company, one of the top four in the global telecom industry." Some Huawei old-timers still remember the first time he made this claim. One day during lunchtime, people were sitting around eating when Ren rushed out of the kitchen wearing an apron and waving a spatula about, shouting excitedly about his prediction of Huawei's future "in just 20 years!" Hundreds of employees were stunned at first, then excited, and then they all began to clap. Half a year later, the story changed from "top four" to "top three" when he brushed off Huawei's competitors in Japan.

Those were the years when people motivated each other at Huawei by boosting each other's courage—a great time to build confidence. In the first 10 years, around 9 PM every night, Ren would go over to the shabby R&D office, a big mug of tea in hand, and call everyone together to sit around, talk tall, and dream big: "You guys are great! What you're doing is greater than anything at Bell Labs." It was clear to the young engineers in their twenties that Bell Labs was the US "heart of science and technology," but they were nevertheless galvanized by the energy in the room. After each pep talk, everyone would feel as if they were pumped full of inexhaustible energy and then they would go back to their overtime, fighting the good fight. Sometimes, in the middle of the night, Ren would take people out for a late-night snack—and yet another pep talk.

As predicted, 20 years later, Bell Labs was in a state of decline whereas Huawei had become one of the world's 100 most innovative companies.

As the saying in Chinese goes, "If you've got the nerve to work the field, in the end the land will yield." For many years after 1993, Huawei's sales volume doubled every year. Every time they set the new sales target for the following year, it was generally recognized as impossible to achieve, and yet year after year, they continued to experience breakthroughs. Hu Houkun, who led sales for many years, said, "Every year we set irrational goals, and every year we were able to make them happen." Therefore, every year at the annual year-end sales conference, thousands of people would stand up and sing the old song "True Hero."

What is organizational vision? It's essentially an organization's traction. Why were so many outstanding young intellectuals willing to follow Ren? Why were so many untamed heroes willing to gather under the banner of Huawei to march towards the same goal? It's because the company's vision was broad enough to accommodate their personal ambitions and sense of mission.

How was this vision generated?

Through dreaming big and talking bigger. When the first person spoke up, the group was skeptical. However, after repeating it a few times, some people started to believe it. And after they kept talking about it more, the number of believers grew, and so on. After 20 years of tirelessly saying the same thing over and over, more than 100,000 followers were soon sucked into the halo effect of Ren Zhengfei's dream. In the end, all the BS—or what we call the "vision"—became a shared feeling within the organization.

By nature, young people tend to be more susceptible to collective unconscious. However, when shattering the illusion of Utopia also serves to shatter their enthusiasm, or when their intensity fades at the broken promise of life's mirages, then they can easily lean towards collective rebellion and individualism. Therefore, an organization's vision must be grand enough to ignite the hearts of the youth; meanwhile, the organization has to put its vision into action in order to inspire individual dedication. Only by aligning the dedication of individuals to the interests of the community can an organization convert its tall talk into something real, make the impossible concrete.

Huawei's former Party secretary, Chen Zhufang, joined Huawei after retiring from her position as a researcher at Huazhong University of Science and Technology and the executive deputy director of its business school. Shortly after she joined Huawei, she said to Xu Zhijun:[4] "I'm a real worker here. You all have stock, so you guys are the bosses."

---

[4] Alongside Guo Ping and Hu Houkun, Xu Zhijun is one of Huawei's three rotating CEOs.

Xu replied, "Ms. Chen, all we've got is a few pieces of paper. What did you think they were?"

For the longest time, many employees didn't care about the value of their equity shares. The 2002 stock allotment required individuals to get bank loans to buy shares. Someone asked Chen Zhufang, "Should I buy some or not?" Chen Zhufang replied, "If you take your money elsewhere, isn't it also a gamble? You might as well bet on Ren Zhengfei."

Around 2001, UT Starcom tried to poach Yu Chengdong[5] from Huawei, promising a high salary and extremely valuable stock options. Yu turned the offer down politely. His "Huawei tattoo," so to speak, went far too deep: He was a master BS artist on the one hand, a master fighter on the other. In Yu's words, "My boss Ren has a way of painting the map nice and pretty; once the map is ready, everyone jumps on board. We all learned this from our boss: Unite people with a ray of hope, not with money."

Yu Chengdong's nicknames are "Big-Mouth Yu" and "Madman Yu." Even among the mid- and senior-level managers at Huawei, a radical and wild group, Yu was considered to be on the extreme end of things, ready to challenge the impossible at any time. Ren Zhengfei used to say: "Let Madman Yu keep his black-and-white approach. The rest of us can adopt a less extreme approach to round things off a bit."

Hou Jinlong was recruited by Yu in 1996 and became the founder of Huawei's GSM business. In 1998, Yu started putting together another team to begin the pre-research stage for 3G technology. Huawei's wireless product line, in typical Huawei fashion, expanded both rapidly and wildly. Over the course of 10 years, led by Xu Zhijun, Yu Chengdong, Wan Biao, and Wang Tao, the team grew from just a few engineers to the tens of thousands of people working there today, accounting for almost half of the company's sales revenue and profit. However, they also went through a period of no income for 7–8 years and they have to take loans from the company in order to afford employee bonuses. What did it take to sustain cohesion in the team? Ambition and vision, together with what they call the "king's mentality."

---

[5] Yu Chengdong is currently the CEO of Huawei's Consumer Business Group, which produces Huawei's line of consumer products, most notably its smartphones.

"I want my team to have a king's mentality," Yu said. "Whatever we do, we want to be the best—able to surpass any competitor." The team would often get worked up to the point of shouting out loud in enthusiasm. As a result of their efforts, WCDMA was the very first product at Huawei that was sold in developed countries overseas first, and then from Europe, its market spread to the rest of the world, including China. Over the past three years, since Yu Chengdong took charge of the company's line of device products, Huawei's mobile-phone business has begun to gain traction in the global market.

What are the company's ambitions for its device business? Yu's "tall talk," so to speak, both internally and externally has been:

> We aim to become the world's leading and most profitable company. We'll go all the way up, despite the temporary setbacks and torment we've gone through over the past couple of years. In the future, we will expand with explosive force, and for that we are storing our strength and building core competences in every domain. Without the ability to look ahead, we are not likely to go far.

The Dialectics of the Turnip and the Pit

Heroes are forged through battle and they're inspired by circumstance; more than that, they go through this crucible on stage. And if the stage is too cramped and narrow, how could its players possibly have enough room to play their part? How can it serve as a gathering place for the best talent? All companies face these questions, and one of the core missions of an organization is therefore to continuously expand, to meet and create customer needs. This is the only way that an organization can realize its vision and ensure healthy development. Xu Zhijun explained why Huawei has expanded beyond the operator business into enterprise networking: "It's very simple—without expansion, what do we do with so many employees?"

Reading between the lines, what he was really referring to was the congestion of power, which can lead to internal strife. The law of conservation of energy has it that energy has to be consumed. If it is consumed for the sake of customers, an organization will thrive; if it is consumed in the course of internal power struggle, chaos will follow.

"One turnip for one pit" is a vivid way that Chinese people use to describe human resource management. Organizational management, in essence, is about continuously digging more pits (jobs), finding more turnips (talent), and making sure the pits and turnips fit. Several turnips scrambling for a single pit would undoubtedly bring about an internal struggle for power.

Obviously, the negative impact of "one turnip for one pit" can be terrible to behold. Cliques may form as a result and laziness can also spread through the organization. For a long period in the beginning and developmental stages of an organization, there are usually more pits than turnips. The HR department and even the organization's leaders have to spend a lot of time finding turnips to fill the pits. At the same time, they have to prevent clique formation and idleness by not allowing the turnips to become too comfortable in their own pit. Moderate and healthy competition can be used as a preventive measure. The lowest-performing 10 percent of Huawei's entry- and mid-level managers are eliminated every year, and members of the board of directors are reelected every five years through a shareholders' meeting. These are important measures to stimulate vitality and prevent people from becoming too hardened in their ways.

Can a small turnip occupy a big pit? This actually involves a question of principle: Does the organization dare to appoint managers through a process of trial and error? In its early years, Huawei was very radical in its leadership appointment. Most young people in their twenties then were promoted twice, thrice, or even five times in a single year: The 26-year-old Zheng Baoyong became the company's second-in-command; the 27-year-old Li Yinan was promoted to director of the Central Research Institute and executive vice president after having joined Huawei only a year ago; some engineers who joined Huawei less than three months or half a year ago were pressured to take charge of an entire project. The company was extremely hands-off, providing money and a platform, then sending people off to do their own thing. Consequently, countless small turnips grew up extremely fast in their big pits. In less than 20 years, many have become global technology experts or managerial leaders in their forties. For example, the president of Huawei's fixed network product line, Zha Jun, is only 44 years old, but Huawei employees call him "Ol' Zha" because he joined Huawei in 1997 and now runs an R&D team of 14,000 people.

Wang Haijie recalled that in 2003, after negotiations with Huawei, Motorola gave up its own wireless product development and opted to bring on Huawei as its OEM. After the negotiation, the two parties went out to a bar for a drink. The white-haired tech leader of Motorola's negotiation team, a chief engineer for GSM R&D, asked Wang Haijie, "How old are you, Jack?" Wang replied that he was 32. The older scientist said in dismay, "When I joined Motorola, you weren't even born yet. I've been doing wireless R&D for 30 years, and I never imagined that we would be beaten by a company that's been in this field less than 10 years." He started sobbing at the bar.

Ten years later, Wang Haijie became the director of the Shanghai Research Institute, managing an R&D team of 10,000 people.

"You either make it or break it. This is Huawei's staffing standard."—In Bai Zhidong's eyes, this is the natural choice for any business. The stage is big enough, but if you don't have what it takes to grow, or harbor disloyalty, you will naturally face elimination.

Risk is the price to pay for being adventurous. Over the years, Huawei made a lot of mistakes in talent appointment, which as a result required the company to establish standardized systems for the selection, deployment, retention, development, and management of leaders. However, the system it has developed over the past decade has become more and more procedure-driven, which has buried a few people with great potential and is not conducive to attracting world-class talent.

Which came first, the chicken or the egg? On the topic of this ancient conundrum, Huawei's executives had a heated debate and came to a final consensus: the chicken. Now, assuming that the chicken came first, how can you ensure that the chicken can lay eggs? And if it can, how many eggs can it produce and what kind of eggs? (Golden eggs? Rotten ones?) These should all be tested for in actual practice. In the worst-case scenario, you'll have a few chickens that can't lay eggs, but so what? Just make a few adjustments. As long as Huawei is open-minded in its human resource management philosophy, it naturally follows that it will be able to attract more golden-egg-laying chickens those who will contribute to the company's prosperity.

In the mid-1990s, annual salaries at Huawei were the highest in the nation. Some state-owned enterprises and government departments voiced complaints. The Ministry of Education specifically called a meeting about this, where many people blamed Huawei for raising the price of labor. A Huawei executive privately responded to the accusation thus: By paying intellectuals a high salary, Huawei can make them stay in China. It's like raising a fat chicken at home to lay eggs, then selling the eggs abroad at a high price. If they can't find what they're looking for domestically and they all leave for Western countries to lay their eggs over there, the Westerners can then sell their eggs back to China at a higher price. Which is better for the country?

Over the years, Huawei has attracted a lot of foreign chickens to lay their eggs in China too, and then it has sold those eggs all around the world. This undoubtedly has a lot to do with an open and radical human-resources culture that doesn't shy away from trial and error.

Back to the subject of turnips and pits, what if you put a big turnip in a small pit? Would that work? In Huawei, it's not uncommon for a senior executive to suddenly get appointed as the general manager of a rep office, a regional president, or the head of some entry-level department. President of sales Mao Shengjiang became the GM of the Shandong rep office; Xu Zhijun, back when he was vice president, became the GM of the Shanghai office; the former president of the mobile device department, Wan Biao, was transferred to Russia as the regional president; executive vice president Xu Wenwei was sent to Western Europe as the regional president; and the list goes on. Many who used to manage thousands or even 10,000 people were later assigned to lead just a handful of people, at most a few hundred.

"Why do we toss managers around like this?" Some people did not understand, some cried, and some even left the company. But in Huawei, this is a very important test: A real phoenix always rises from the ashes. A manager should be able to go through ups and downs, move from one pit to another and thrive, and grow in what the company calls a "zigzag" way. Most of the senior executives in Huawei have served only one to two employers (either companies or government agencies) after school, having spent their entire youth at Huawei, but they nevertheless gained experience in 5–10 different positions—sometimes even more—within Huawei itself. One of them was in 15 different departments in 23 years; Tian Feng, a member of Huawei's board of supervisors, held more than 10 positions. Such rotation not only gives full exposure to a person's potential, it also tempers their will.

However, the strategy of stuffing big turnips into small pits has another layer of intent: Although the current pit is small, it might be of strategic importance in the future. High-level commanders can use their vision, influence, and ability to organize resources and quickly expand the pit. As a result, the team, the market, customer penetration, sales, and profits will also be able to grow more quickly.

Here, the same "make or break" principle also applies. If a big commander can't achieve a rapid expansion target or makes little contribution, he or she will also be held responsible and punished, including not making it back to their previously high position.

What is the epitome of fairness within a company? Using a "horse–racing" system for talent management. Horses gain recognition by racing well. And only the best racehorses will bear their organization's banner boldly and in the right direction.

The Enemy of Innovation: Incompatibility between Black and White

The Vast Inner World of a Leader

A decade ago, Ren Zhengfei read an article written in remembrance of Mao Zedong, the founding Chairman of the People's Republic of China. He was impressed by what Chairman Mao had told his assistants: There are no fish in an absolutely clean and clear river. Ren Zhengfei agreed. He also believed that a river, in the real sense, refers to a torrent that carries mud and dirt along with it. Literally speaking, leaders should possess open and liberal minds. Their duty is to put a sound system of rules and mechanisms in place, or to build up a dam, so to speak. The dam should be solid and high enough, and the course of the river should be wide and deep, so river torrents can be allowed to flow through quickly. At the same time, there should be gates where the water can be let out in the event of a flood.

Similarly, some believe that a leader should have a vast inner world with a waste-recycling facility that can collect, sort, and process any waste. All organizations, after all, are spaces full of conflicting information, conflicting personalities. As the author Yin Luobi wrote in an article titled "A Private City under Water:"

> The psychological troubles that people suffer from are mostly caused by spending too much time with other humans. Depression, anxiety, obsessive compulsiveness, schizophrenia, delusions—they all mostly stem from a single reason, that one big question: who am I in the eyes of others?

Therefore, it's a leader's duty to create a proper evaluation system that can realign personal pursuit with the goal of the organization. At the same time, such a system should both embrace personality differences and keep personal sentiment from getting in the way of others' interests (and the interests of the organization itself). Another critical factor that affects the well-being of an organization is the health of its subcultures, where the core problem is the spread of rumors or unverified information.

Some entrepreneurs use extreme examples: "Instead of pissing all over the place along with them, you'd be better off building toilets." Where should the "toilets" be built? In the inner worlds of the leaders and managers. Employees must have a regular channel to freely express their opinions, whether they are wrong, false, hostile, or radical. If a leader can't accommodate and digest differing opinions, including the ugly and dirty ones, then they will end up overflowing into the bigger community.

Because of this, the analogy of being a waste-recycling plant is perhaps more appropriate. Leaders must listen to the voices of their team, but they shouldn't deceive themselves in the process. In the words of Ren Zhengfei, "Whatever you resist will only get bigger."

Leaders must possess soft ears but sharp eyes. They must be able to sort through and ascertain what's constructive, what's useless, what's harmful, and what's radioactive or erosive. Accordingly, they should bury what needs buried, incinerate some things, let others biodegrade, and recycle whatever is useful. This process helps clean up the environment of the organization and enhance its power. To achieve this, a leader must possess a clear mind, an open heart, and a firm hand.

There was once a flood of attacks on the internet against Huawei and its leaders, and it was clear that Huawei employees were behind them. There were complaints, feelings of discontent, rumors, and "uncovered secrets." How should the company have responded? After intense debate, senior management reached an agreement based on the concepts of openness, compromise, and grayness. Ren Zhengfei said:

> Each man is born with a mouth, so he must be allowed to speak. Of course, if too many people speak at the same time, it would be too noisy. This is inevitable. We can't stop people from speaking or punish anyone who speaks. If we do that, we would only fan the flames of hostility.

Therefore, Huawei decided to set up a web forum, the Xinsheng Community (literally meaning "the voice within"), where its employees are encouraged to voice their opinions on the company's regulations, policies, and decisions, and can further engage in free debate without any interference. Several years later, this web community, which was intended merely to increase transparency and appease employee discontent, evolved into an important communication platform for senior management.

In January 2012, the opinions on this forum were collected and organized in a booklet entitled *Huawei's Issues,* which was then distributed to each executive.

Speaking of the forum, Ren Zhengfei said:

> We were taking a risk when we decided to set up this online forum, but surely we made the right decision. This proves that

we can't attempt to lock anyone's mind or shut anyone's mouth. The sky will not fall if we let people speak their minds. Instead, if they speak out fully, they can help mend the sky when it leaks. I believe that every person has their own goodness, and wisdom comes of pooling all their goodness together. I also believe that everyone has their own insight, and we can make great decisions if we pool all that insight together too.

It's critical to note that leaders should not attempt to monopolize power, control information, or hold the exclusive right to interpret information. On the contrary, leaders must be open-minded and tolerate differences.

The Source and the Stream: Tolerance Is a Type of Strength

"My Father and Mother" is a popular essay that has been translated into dozens of languages and has struck a chord with many people. However, in the eyes of Ren Shulu (Ren Zhengfei's younger brother), "It's a plain article. He was away at school, so he didn't really know much about all the things our parents suffered." One scene he depicted was of their father, who was labeled as a "capitalist roader"[6] at that time. One day, his father came home with smile on his face and kept telling their mother, "Someone said hi to me just now, and it was really nice, really nice." Imagine this senior scholar who, because of so-called "historical issues," was bullied and alienated by society, isolated from his colleagues and students to such an extent that when he received a simple greeting from a stranger on the street, he felt so greatly flattered that even his long-lost smile found its way back to his face.

In fact, feeling cornered by others' ignorance and insularity was not foreign to Ren Zhengfei when he was a young man. Therefore, after he set up his own kingdom at Huawei, he strived to build what he cherished most—an inclusive and tolerant corporate culture.

---

[6] "Capitalist roader" refers to a person or group of people that harbored capitalist ideals during the Cultural Revolution—those who proactively "chose the road," so to speak, to capitalism.

Ren Zhengfei once gave a vivid speech about tolerance:

> It's the key to a leader's success. Whatever job you do, you need to deal with either people or things. Take a scientist, for example. A scientist might have a strange and eccentric personality, but if her job is to deal with instruments in the lab, then a lack of tolerance isn't a huge deal. Similarly, if a worker can operate his machine very well and produce perfect products, even though he can't get along well with other people, it matters little. Corporate management, however, is a completely different matter. All managers have to deal with people, and management is defined as "a skill to accomplish a task through the hands of other people." The moment you deal with people, you'll see the importance of tolerance.

In recent years, Huawei has deliberately encouraged and disseminated opposing voices within the company. It has even set up a "Blue Team" that serves as an opposition to the "Red Team" (the company's principal force). [7] Outstanding performers from the Blue Team are appointed as the commanders of the Red Team. Ren has said that "only those who can find faults with the company will be able to make it better."

Bai Zhidong, the founder of Huawei's Access Network, was once the GM of the company's Shanghai office and later the head of the pricing center. Wherever he went to work, he never failed to demonstrate his "non-conformist" or "heretical" behavior and opinions. As a straightforward man of unrestrained personality, he really missed the Huawei of the 1990s: "Nobody would interfere, so you could do whatever you wanted, as long as you delivered in the end." When I asked him how someone with his personality was able to last in Huawei for 20 years, he answered, "Because I was tolerated. Huawei's culture, on the whole, is accepting." When Huawei was building up its Blue Team, Bai Zhidong was recommended by Xu Zhijun to be the commander.

---

[7] As noted in previous chapters, Huawei employs a "Red Team and Blue Team" method of self-reflection to identify potential weaknesses in the company. Similar to modern war games, the Blue Team plays the role of a competitor, always challenging the Red Team and exposing its faults to help the Red Team improve.

On the topic of the Red and Blue Team arrangement, Ren Zhengfei said the following:

> The Blue Team is everywhere in the company. Every part of the organization includes members of the Blue Team; it doesn't exist only at the top level of the organization. Our minds are also composed of red and blue, and I believe people are born with red and blue inclinations. I have always attempted to deny my own will, and I have always been critical of my own decisions. I believe the Blue Team exists in every field and every process, and the opposition between blue and red is always present. I think I can accept oppositional forces in Huawei. We have to accept and unite everyone in the organization, including those who are opposed to our policies. Even if they from an oppositional alliance in the company, it doesn't matter as long as they are not trying to provoke dissension with malicious intent, but to raise divergent ideas on technology. If we allow for more diversity in the company, I believe we can motivate wisdom and talent. Conflicting opinions are part of our strategic reserve.

In the meantime, the assessment Yu Chengdong received from his tutor at graduate school was: "He is not fit for working in a company or with others, but can be a domain expert." Yu saw himself as a catfish belonging to "the unconstrained, the radical, and the restless; not especially endearing to leadership, someone who goes against the spirit of harmony and the doctrine of the mean in Chinese culture." When he was fresh off campus, Huawei gave him the impression of a peculiar place, filled with people who failed to graduate from school, who were dismissed by their previous employers, and who gave up their previous jobs to join the company in secret. Many employees did not have formal positions at Huawei and the working atmosphere was absolutely free—even video games were allowed during working hours. In the office, you could find female employees practicing handstands against the wall and people's feet up on the computer keyboards. And yet everyone was working like crazy, more than 10 hours every day. Management was chaotic at best and the computers were not locked up. Once someone nabbed a hard disk, copied all of the company's R&D materials, and jumped ship to a competitor.

Yu Chengdong witnessed the entire process of Huawei's evolution from chaos to standardized management. In a sense, it's only natural for standardization to round off people's rough edges and suffocate individuality. Yet this does not seem to be true for Yu Chengdong, who is still unapologetically himself. He is perhaps one of the most-criticized senior executives at Huawei, but according to him, "Although our boss is short-tempered, he is very kind at heart. Huawei is a very tolerant company, which has allowed me to survive."

In truth, Ren Zhengfei doesn't hold leaders like General George Patton or Li Yunlong[8] in great esteem. He has more admiration for Eisenhower, a leader who transformed himself from an army general to a strategist and then to a head of state. When Yu Chengdong was assigned to serve as the president of Huawei Europe, Ren Zhengfei said to him: "Now you need to learn how to be a leader."

At Huawei, most managers are not in fact leaders, but rather army generals or soldiers with their own respective strengths, weaknesses, and striking personalities. How does Huawei deal with this group of people, who comprise a majority of the company's workforce? Furthermore, how does Huawei deal with managers or employees who made mistakes in the past?

The day Steve Jobs passed away, Ren Zhengfei was with his family in Lijiang, Yunnan. When the news came, his youngest daughter, a fan of Steve Jobs, proposed that they pay tribute to the deceased, so they honored him with a moment of silence in the outdoor café of the hotel. Ten days later, Ren Zhengfei said:

> China doesn't have the right soil to nurture another Steve Jobs. We are not tolerant enough, and we do not protect intellectual property well enough. Millions of Chinese mourned the death of Steve Jobs, and yet we still can't seem to offer the same tolerance for China's own entrepreneurs. Innovation is only

---

[8] Li Yunlong (1919–1965), a major general in the Red Army. He is known for his extreme charisma and for his ability to get anything done if he set his mind to it. He is also known for setting his mind to things that *he himself* wanted to do, not necessarily what he was ordered to do.

possible with tolerance. And the same is true of cultivating great business people.

The connotations of what he said are multifaceted: A nation that does not tolerate mistakes and failure will suffocate its creativity and its elite talent. Steve Jobs had an eccentric personality, being an unworldly genius who smoked marijuana in his early years and whom American society accepted with great tolerance. Tourists visiting the United Kingdom will come across a lot of statues along the streets: They include not only great thinkers, artists, and kings, but also infamous tyrants and clowns. China, however, remains firmly rooted in a black-and-white duality. For example, Chinese soldiers who had been taken prisoner in the Korean War 60 years ago were subjected to cruel torture on the charge that they had surrendered to the enemy. In contrast, captured US soldiers have returned to their country as heroes.

Over the past decade, Ren Zhengfei has shared the same story whenever he gets the chance. The story is about two people: Wernher von Braun, a former Waffen-SS official during World War II who later became the father of US space travel, and Sergei Korolev, the father of the Soviet space program who opposed Stalin. These two space heroes had both been under the death penalty, but they were granted the opportunity to redeem themselves by their respective countries and later made outstanding contributions, with diametrically opposed ideologies. As a result, the Space Race between the United States and the Union of Soviet Socialist Republics (USSR) in the early days was literally a race between two condemned prisoners.

Alan Turing, the father of modern computers, was a firm believer that machines could ultimately think like human beings. Yet his valiant and tragic life belies the notion that there is no difference between the human mind and machines. As a living, breathing compound of blood and flesh, Turing was far more fragile and complicated than a machine. On June 7, 1954, Turing took his own life at the age of 41—falling victim to the intolerance and cultural prejudice in British society—because he was homosexual.

Turing's genius ushered mankind into the Information Age. For this reason, Huawei named one of its laboratories the Turing Lab.

In the words of an English proverb, "To err is human, to forgive is divine." An ancient Chinese writer said essentially the same thing,[9] and this idea was broadly accepted early on in Chinese history. But in modern mainstream culture, yes is yes and no is no. Black and white do not mix. But the truth is that nobody is perfect: If you are unwilling to accept a person as they are, you will find yourself alone in this world.

A lot of Chinese companies were quick to rise and quick to fall, and one of the major reasons behind this is the narrow-mindedness and intolerance of their founders and management teams.

In late 2013, the board of Huawei passed a resolution: "To Unite All Forces that can be United."

In June 2014, Guo Ping, deputy chairman and rotating CEO of Huawei, met the Irish Prime Minister Enda Kenny, who asked him why Huawei chose to set up a research institute in Cork, which is 280km away from Dublin, the capital of the country. Guo Ping answered: Because in Cork, there is an expert in network architecture who's unwilling to leave his hometown. For his sake, Huawei decided to set up an R&D team right where he lived.

At the core of Huawei's talent policy is an agree-to-disagree attitude, a recognition of differences, a commitment to objectives, a focus on the job at hand, and a platform that lets people all over the world put their talents to use. According to the president of a multinational company, Huawei is the only company in the world today that has managed to fully utilize both Chinese and non-Chinese talent. Among the sixteen Fellows at Huawei, nine scientists are of foreign nationality and seven are Chinese.

If a huge amount of money is spent on a fancy vase that that looks nice but can't hold water, it's agony for the vase itself and for the person who bought it. Huawei naturally tries to avoid this when seeking out high-end talent. In history, military advisor Xu Shu was hired by Cao Cao's[10] camp,

---

[9] Here the author is referring to the famous saying "Nobody is a saint. Who doesn't make mistakes?" in the *Zuo Zhuan,* a widely read and frequently quoted commentary on political and military affairs during a period of the Zhou Dynasty from 722 BC to 481 BC.
[10] Cao Cao was a major warlord and king during the Three Kingdoms period of Chinese history.

where he enjoyed the entitlement of his position but didn't contribute to the war cause. It was a loss to Cao Cao's camp and a sad experience for Xu Shu as well. Huawei seeks to integrate the commercialization of scientists' ideas with a "marriage" between scientists and engineers, ultimately taking the results of their research and development and giving it practical application in the market. In a word, Huawei seeks to retain people by treating them well and to motivate them by providing development opportunities.

Leaders Must Surrender Their Imperial Mentality

On New Year's Day, 2012, Ren Zhengfei's speech to the executive management team, "The Spring River Flows East," was published on Huawei's intranet. Within three days, there were 600,000 views. There was also a big stir among domestic media. They wondered: Is this man the mysterious iron man we used to know? Is this the real Ren Zhengfei?

Yes, of course. This was the real Ren Zhengfei, a mixture of pride, confidence, fear, and loneliness. In the speech, he gave an account of his personal experience. In his boyhood, he had heard the story of Hercules from his mother and also read a great deal on Chinese legends. Back then, he worshipped powerful heroes like Li Yuanba and Yuwen Chengdu[11] of the Tang Dynasty. He also believed in the absurd tale of Zhang Fei fighting against Yue Fei.[12] Later, in his adolescence, he read about Li Qingzhao, the female poet from the Southern Song Dynasty, and assumed with adolescent sensitivity that the poet was in love with Xiang Yu, one of the all-time heroes in the history of China. She wrote in one of her poems, "Alive, be a man of men; dead, be a soul of souls." This became the motto of many in those days, and dreams of heroes inspired young Chinese people to keep up the struggle, to study hard, and excel at school.

---

[11] Li Yuanba and Yuwen Chengdu are legendary heroes who were first written about in the novel *Shuo Tang,* a collection of heroic stories from the Qing Dynasty. They rank among the top 18 warriors in China (with Li Yuanba ranked number one), and are presumably based on real men in history.

[12] Zhang Fei and Yue Fei are two well-known generals that existed at different times in Chinese history, hence the absurdity of them having fought with one another. This might be akin to reading a story about Dwight D. Eisenhower's epic fight with Robert E. Lee.

And it's because of this hero worship that Ren Zhengfei met with one setback after another. In primary school and high school, he was a solitary student and was denied acceptance into the Communist Youth League. In the PLA, he did not become a CPC member until very late in his military service. Life before he turned 40 was solitary, lonely, and hard.

Ren Zhengfei, who had never led anyone before, founded Huawei in his forties. Much like those historical figures who rose from nothing to take on the world, during the early years of his entrepreneurship, he gathered a number of brothers around him who shared in weal and woe. They were not separated by positions in the company; they divided the gains equally among themselves, like the *Outlaws of the Marsh*. Ren said:

> I founded Huawei by myself, and in China it was called a self-employed business. In those days, to organize a big team was wildly presumptuous—the timing wasn't right. I designed the employee shareholding scheme soon after I founded Huawei. I had intended to unite all my colleagues by a certain means of benefit sharing. At that time I had no idea about stock options. I didn't know that this had been a popular form of incentive for employees in the West, and that there were a lot of variations. All the things I went through in my life made me feel that I had to share both responsibilities and benefits with my colleagues.
>
> In Huawei's early years, I had left our "guerrilla commanders" alone in managing business operations. As a matter of fact, I didn't really have the ability to lead them. During the first decade, we rarely had any operational meetings. I flew to different parts of the country to hear their reports, tried to understand their situations, and gave them the "go ahead." I listened to the brainstorming of the R&D staff. R&D was a mess at the time. We had no clear direction—we were like a bunch of flies in a bell jar, bouncing around, flitting about. The minute a customer demanded improvement, we would go to great lengths to meet that demand. Financial management was an even bigger challenge because I don't understand finance at all. In the end, I had not managed the relationship with finance staff well and, to my regret, promotions for them were rare. Perhaps it's because I wasn't capable, because

I was foolish. I had given so many people in the company the freedom to express their brilliance to its full extent, which brought so much success to Huawei.

Back then I was called a "hands-off boss." I had wanted to be hands on, but I didn't know how. Around 1997, there appeared a new group of feudal lords in our midst, each with their own set of ideas and intentions; internally, our philosophy was chaotic, vastly different modes of thoughts loomed above us like a thick forest canopy. No one knew where the company was headed, so I had no choice but to lead….

In such plain and sincere words, Ren Zhengfei described the early history of Huawei. Some scholars believe that this is common in the history of China. Most social or political organizations in China were founded with plain equalitarian culture and simple democratic underpinnings, such as the Taiping Heavenly Kingdom and the Tongmenghui (Chinese Revolutionary Alliance) in the late Qing Dynasty. But when they grew too big, their top leader began to monopolize all forms of power, both big and small. Everyone in the organization would begin to follow his lead and any form of dissent or suggestion would be shot down on the spot. This phenomenon seems to prevail in many Chinese organizations today: even universities, research institutes, and trade associations are no exception.

The imperial mentality is deeply rooted in China, an ancient country with a 1,000-year feudal tradition. It is also very common in China's businesses. One might be a common person, like a farmer, teacher, or low-ranking government official, when starting up a company, but after time passes and the company secures its own solid tract of business territory, its leader will suddenly gain an inflated sense of self and become as domineering as an emperor. Cult of personality then takes hold within the company. The founder might start to believe that the company's success is owing to his or her personal fortune and forget about the endeavors of the group throughout the years.

The tragedy is that when an entrepreneur becomes an emperor, the company loses its grasp on grayness. White and black become two completely separated things and the standard for telling white and black apart becomes subject to the will of the big kahuna.

When it is controlled by one leader, one voice, or one idea, an organization will turn rigid and vulnerable. And then collapse is just around the corner.

Ren Zhengfei is the unquestionable leader of Huawei, but he is not an emperor. To create sensation, the media call strong leaders "kings" or "emperors", but this is usually misleading and biased.

Emperors and kings are those who stand far above their subjects and are detached from the real world. They live, or rather are imprisoned, in an enclosure. His or Her Majesty is beyond any challenge: Whatever they say is the new golden rule. The empire or kingdom under their rule, therefore, is strictly governed, a place where black and white do not converge and the atmosphere is intense, if not suffocating.

When Steve Jobs was given the crown of a tech emperor, it was a misunderstanding and an act of blasphemy, not one of true extolment. Aware of this, Ren Zhengfei has been trying to tear away the Emperor's New Clothes that he himself has donned; he says that he knows only a little bit about everything and nothing about anything—that he's a cultural instructor and that's it. He said:

> I am half-literate in subjects like technology, corporate management, and financial affairs. I am trying to pick up and learn about these things along the way. So I must modestly gather a group of people, and manage their collective strength to drive the company forward.

At a recent board meeting for peer critique and self-criticism, someone proposed a list of questions to evaluate the boss: Is Ren Zhengfei, our boss, versed in technology? Seven of the attendees said no. Does the boss understand sales? Again, seven said no. Is the boss an expert in corporate management? One of them said no, and that man was Huawei's deputy chairman, Xu Zhijun. Xu is one of three rotating CEOs in the company and he is energetic, straightforward, and sensitive to business trends. Some liken him to a clever fox with a keen nose, able to detect new opportunities before others do. Moreover, he understands the importance of cooperation. At a meeting with Xiao Gang, chairman of the Bank of China, Xu Zhijun said that IPD and ISC represent a huge and far-reaching transformation of Huawei's corporate management system and that Ren knows nothing but the literal meaning of the two terms. Nevertheless, Ren has stood firm behind the transformation program and has put the right people in place to manage it. Through this massive 14-year program, Huawei achieved its current standards of R&D competence, its end-to-end supply process, and its ability to serve its customers worldwide.

Xu Zhijun has also, in some sense, torn away the façade of the Emperor's New Clothes on Ren Zhengfei, telling the true story of a man who is unique and able to step back in areas where he is not well versed. Yes, Ren Zhengfei is half-literate in technology, sales, and corporate management. But he is, in fact, a corporate-management philosopher. To put it in Ren's own words, "Over the past 20 years, I have been dealing with the abstract. More precisely, I have spent 70 percent of my time on abstract matters, and only 30 percent on concrete issues." His task is to learn, contemplate, exchange, and communicate ideas.

Traumatic Memories of Teenage Years: Hunger and Sharing

Since 2012, Huawei has managed to shed its iniquitous and demonized reputation around the world. From its increased openness and transparency, people came to realize that, behind the success of this Chinese high-tech company, there has been a convincing trajectory of development and a fleshed-out philosophical system. In other words, the Huawei story is a brief history of group struggle and wealth creation, spurred forward by values that reflect human desire—a story filled with joy and suffering, hope and despair. At the same time, it is also an exploratory history of thoughts on managing human nature.

When I was a visiting scholar at the Business School of the National University of Singapore (NUS), more than one professor probed with great curiosity: Huawei's management philosophy and practices are clearly something that contemporary Eastern and Western management courses can't cover in full. Too many aspects go beyond existing frameworks and models, or are even original concepts developed by Huawei, like the reform measures that led to two massive waves of employee resignation. So the question is, what is the source of Huawei's management philosophy and where it is heading?

My answer was: Go and look for them in "My Father and Mother." Written in 2001, this colloquial essay is a very candid monologue by the founder of Huawei, Ren Zhengfei. It's not only a memorial essay about his parents, but a confession to himself as well as a statement to his employees, to the general public, and to those who have shown either care or expressed doubt about Huawei. The origin of all the ideas and practices that Huawei has embraced during its 27 years of development can be traced back to this one essay.

The pangs of hunger are an undercurrent throughout the essay. Looking back as a 56-year-old man, when he wrote about his experience, his deepest impression from the Great Famine in the 1960s was of a piece of corn pancake that his mother stuffed into his hand each morning. It was "saved from the mouths of my parents and my younger siblings." At the end of each month, his mom would ask around to borrow CNY2–3 to save this family of nine from running out of food. Two or three family members shared one quilt and the bedsheets were laid out on piles of straw. He never had a thin undershirt to wear during the hot summer, only thick clothes, because they didn't have different clothes for different seasons.

The feeling of hunger is a memory held in common by this generation of people that experienced long periods of starvation and it left many of them with traumatic memories from their youth. No matter if they're deep or shallow, abundant or sparse, as these memories began to settle, they undoubtedly had an impact on the personalities and mindsets of the people who grew up in that environment. Based on observation and unempirical judgment, scholars have come to the hypothetical conclusion that trauma caused by hunger has left the generation born between the 1940s and the 1960s in China with a shared fear of scarcity. And as soon as an opportunity to release such fear or anxiety comes along, it generally breeds greed and an excessive desire for tangible wealth, as well as intangible power and fame. The table manners of Chinese people who are currently in their fifties, sixties, or seventies are a good example: First, they tend to order a lot of food, regardless of how many people are going to consume it; secondly, they tend to shovel down food as fast as they can. Even the way they walk is rushed. You will rarely spot them calmly strolling about.

I chuckled to myself when writing that: I am exactly that the same way. And so is Ren Zhengfei.

Let us assume for the time being that "greed" is a neutral word. If this was the case, it could be regarded as one of the greatest sources of momentum behind progress in human history. Out of greed grows the ambition to change the world. China's economic blastoff over the past three decades has been propelled by the unbearable hunger pangs of two generations and the pursuit of adventure and progress that followed. Hunger is what nourished outstanding leaders and entrepreneurs across all different sectors and what laid the foundation of China's strength as a nation. However, there have also been a large number of risk-takers who have fallen under the spell of insatiable greed.

Without question, Ren Zhengfei is greedy. In 2007, when he was listening to a business performance report by Yan Lida, head of Huawei's rep office in Japan, he said impatiently, "I have no interest in listening to these. What I want is the whole world!" Such an appetite or craving for "the whole world" has underpinned the international success of a number of Chinese companies like Huawei, who all achieved it in less than 30 years.

However, enormous greed is just one aspect of his personality. Another redeeming or even opposing aspect is the restraint he shows towards the desires of individuals and the organization, or simply put, his willingness to share. Huawei's employee shareholding scheme, the concept of "Digging Deep Channels, Building Low Weirs" in operations management, and the concept of collaborating with industry peers to develop a balanced global business ecosystem are all manifestations of his strong sense of rationality and self-control.

Survival is the greatest driver of Huawei's development, and in the past, it has simultaneously served as the organization's most basic and most lofty goal. For Huawei, the only way to survive is the relentless pursuit of progress, which means paying dearly with blood and tears. On the other hand, Huawei's long-term survival is predicated on surviving *together*, which means sharing together, developing together, and achieving prosperity together with its employees, customers, and competitors. If customers go downhill because Huawei sucked them dry, then Huawei will lose all of its reasons for survival; if employees can't fully enjoy the fruits of the company's development and interests are largely tilted in favor of shareholders or a small number of executive members, how can the company survive in a sustainable and healthy manner? A lack of competition, on the other hand, is not only a disaster for the industry, but for the "king of the hill" that dominated the industry. A lack of competition would be the harbinger of its fall from the top of the world to the very depths of the abyss.

Some of the origins of this management philosophy at Huawei can be identified in the article "My Father and Mother." In a way, the tolerance, perseverance, and commitment that have manifested in Ren Zhengfei's management style were shaped by his father's—Ren Moxun's—unwavering beliefs in the face of intense suffering caused by the polarity of "black and white thinking" in society. Shortly after Huawei was founded, Ren Zhengfei discussed the idea of an employee shareholding scheme with his father, who studied economics in the 1930s, and he received great encouragement.

In my eyes, however, the deepest influence on Ren Zhengfei's personality comes from his mother, Cheng Yuanzhao. I actually interviewed Ren Zhengfei's younger siblings about this: From whom did Ren inherit his personality traits? Their unanimous answer was: mom. It seems that many outstanding figures, both in China and elsewhere, owe their success to their ordinary yet remarkable moms. All of these women have nurtured a handful of distinguishing traits in their children during their formative years, such as perseverance, endurance, tolerance, diligence, self-discipline, and self-respect.

As for Cheng Yuanzhao, she instilled in Ren Zhengfei a strongly held belief in group survival—everyone living and surviving together. The idea of an equal system of allocation is based on this belief. At 15, Ren Zhengfei experienced this system at home, where it was aimed at controlling everyone's desires—a "strict system of rationing," to be exact—to "ensure that everyone will survive." In his unadorned essay "My Father and Mother," Ren Zhengfei twice hypothesized that, had it not been for their food allocation system, perhaps "one or two of my siblings wouldn't be alive today." What a harrowing chapter of youth.

We have reason to infer that these memories have shaped the cultural psychology behind Huawei's core values of customer centricity, dedication, and perseverance.

"The highest degree of selfishness is self*less*ness."—Ren Zhengfei has been repeating this sentence for more than 20 years. Everyone has selfish ideas and everyone has desires. This is especially true for people who go against the flow and know how to bide their time, or those who have a more desperate sense of hunger than others. However, for those with a strong sense of mission, with their sights set on the bigger picture, they need to control and balance their desire more carefully: Wealth distribution encourages the group to generate more wealth, creating a cyclical accumulation of more wealth; power distribution, on the other hand, motivates everyone to exert their best possible effort so that power will be exerted to the maximum degree. Even when it comes to fame, Ren Zhengfei often transfers the "proprietary rights" of his opinions to peers or subordinates, because "what I want is success, not fame—fame doesn't put dinner on the table" and "fame is just food for dogs," the latter of which is said to be his parents' motto.

Although we might think that society today in China can be rather materialistic—filled with boundless desire—there are, in fact, still a large number of brilliant people across different fields who have a firm sense of determination and mission. They are able to effectively convert their personal greed into the ambition of a group or an organization and to strike a dynamic balance between desire and moderation. Obviously, if a business organization can't create a big "pie" for its followers to consume—a big enough "China pie" or "world pie"—then it won't be able to sustain their passion to compete, nor will it be able to satisfy their instinctive desire for wealth, power, or fame. As a result, there's simply no way for the organization to achieve sustainable success. Nevertheless, desire should be kept under control, including that of the organization itself, individuals within the organization, and especially the desires of the founder and management team.

After Ren Zhengfei founded Huawei and established the sharing mechanism for dedicated employees, he put the company on a path of no return: the contract-based credit culture. When employees willingly accept the principle that "dedicated contributors will get rich and get promoted," it was the equivalent of giving Ren Zhengfei and all the managers in Huawei a "line of credit," so to speak: Were they really going to follow and implement the principle of value creation, assessment, and distribution for a prolonged period of time? Were they ready to abstain from your desire to maximize individual wealth, power, and fame? Of course, it's not easy to establish a good system, but it's even more challenging to keep a good system in shape and stick to it. If the credit is abused, employees won't stay with the company for long. The minute credit abuse becomes commonplace or severe, the company will collapse. In turn, the company won't allow employees to violate the contract-based credit principle (the credit is bidirectional, after all) by reneging on agreed-upon values.

Dedication—The Foundation of Huawei's Human Resources Management is a book that documents the core management concepts developed by Ren Zhengfei and his management team over the past 20 years. If you read it carefully, you may find to your surprise that, for more than two decades, Huawei's fundamental value chain has remained unchanged. No matter how the internal and external environment has evolved or how the context of Ren Zhengfei's discourse has changed, behind all those differently packaged expressions lies always the same pattern: Customer centricity and dedication are the key to success. This is an agreement between Ren Zhengfei and every employee at Huawei, and also an agreement between the whole management team and every employee at Huawei. It is a cultural agreement based on common trust and belief. No one, especially not senior executives, can betray or distort this agreement; otherwise the company would fall.

Obviously, this concept has been planted into every Huawei employee's DNA and there is no reason for them to get rid of it—fusty old common sense is often the crystallization of human nature throughout history, and is therefore universal truth. When it comes to this one core aspect of the company, Ren Zhengfei is firmly opposed to innovation.

For Ren Zhengfei, this concept comes from the profound life experience that occurred over 50 years ago. For managers at Huawei, one success after another during the past 27 years is also clear proof of the efficacy of this concept, so there is no need to test it repeatedly or question its relevance. On the contrary, it needs to be promoted and communicated over and over again and, in particular, it needs to be integrated more solidly at the system and process levels.

It is worth mentioning that the fundamental purpose of Huawei's core values—customer centricity and dedication—is to guide an organization of more than 170,000 people to charge forward and attack, not to maintain the equilibrium of "surviving together." Equilibrium is only temporary, and excessive equilibrium is detrimental to an organization's long-term development, though it can make the management appear to be perfect and internal relationships appear to be harmonious. The ultimate goal is not unity, but success. However, success can't serve as a reliable guide for the future. In the past, Ren Zhengfei said that survival was Huawei's most basic and lofty goal, but in hindsight, this reveals the limitations of his perception 10 years ago, back when Huawei was confronted by the colossal challenge of survival and development.

"To survive and survive together" was the most base-level objective of Huawei in its early days, which reflected its simple and natural instinct for existence at that time. However, the world has changed drastically and Huawei is no longer what it was 10 years ago, nor will it face the same challenges in the next 20 years. As a result, substantial adjustment has to be made to its most high-level objective—that is, based on the minimal requirement of "surviving together," top performers in the company need to be motivated to move further ahead so that the whole organization will be led by a group of "future stars"[13] into an unexplored strategic territory in the industry.

Only by doing this will Huawei avoid the fate of its predecessors, who ended up failing or subjecting themselves to all manner of problems, whether because they pursued highly balanced management, or because

---

[13] Huawei votes on the "future stars" among its employees. In 2015, more than 30,000 employees were recognized as future stars, accounting for 20 percent of the total workforce.

they carried out defensive innovation to maintain the status quo, or because they just focused on the realistic growth potential right at the tips of their noses. Huawei must hypothesize and innovate with its sights set on the future, which means that its culture has to be largely built on a sense of urgency and aggressiveness.

Huawei Culture: Both Chinese and Western, Neither a Donkey Nor a Horse

So how should we define Huawei's culture? It's neither a donkey nor a horse, and it is both Chinese and Western. Idealism is its banner, utilitarianism its guide, adoption and adaptation its driving principle. In the eyes of the West, Huawei operates based on Eastern logic and plays the game by Western rules. And in the eyes of the East, no one knows whether Huawei is playing Chinese chess or *go*, bridge, or mahjong. In essence, the corporate culture at Huawei is diverse, fuzzy, and gray.

Some management scholars define Huawei as a company ruled with Maoist thought. This, however, is a huge swing and a miss.

Mao Zedong called on his comrades to "stick to the right political direction, adopt a simple and pristine work style, and employ flexible tactics." This is an instruction that Ren Zhengfei has held as gospel, as it fits perfectly with business organizations. No doubt Mao is an organizational master, the likes of whom is rarely seen in history, and no doubt that, deep within, Ren's ideas have a degree of Maoist and communistic residue. This is common among people from his generation and it naturally projects into their life and career. According to Ren Zhengfei, commitment to dedication and valuing dedicated employees—or in other words, more pay for more work—are the very values that Huawei has learned from the CPC. In fact, there are many other such examples: Huawei has held peer critique and self-criticism meetings for more than 20 years, has announced information on managerial appointments for the last decade, and has set up a committee for integrity and ethics compliance.

However, at the core of his philosophy, Ren Zhengfei advocates grayness, or the convergence of black and white, which have traditionally been set against each other. As part of this philosophy, he attaches great importance to openness and compromise. He is opposed to conflict but advocates constructive partnerships, which clearly sets his approach apart from Maoist thought. Therefore, the argument that Ren Zhengfei has been running his company based on Maoism is simply incorrect.

Ren Zhengfei once recommended a book to his colleagues—*Hu Yaobang: Rectifying Wrong Charges*. He had wanted his colleagues to learn to endure unjust allegations and thrive amid defeat and frustration.

In 2004, when the whole country was commemorating the 100th anniversary of the birth of Deng Xiaoping, Huawei sent an email to all employees in the name of its CEO, recommending an article, "Rethinking Deng Xiaoping's Philosophy." Deng Xiaoping's political philosophy was both anti-Left and anti-Right, and had a far-reaching effect on Ren Zhengfei, who drove Huawei forward in balanced waves over the past two decades.

Ren Zhengfei is keenly interested in history and he often goes through different subjects to find one that suits Huawei's development. For instance, he loved the TV series *The Qin Empire* and bought hundreds of copies as an expression of his support for serious forms of historical screenwriting, production, and investment. From this historical narrative, he realized how hard it is to carry out reform. The ability to handle everything throughout the process of reform is an elusive art, including how fast and how forcefully the reform can be pushed forward and where the reformer should pause for respite. It demands a strong grasp on grayness in order to win support and limit resistance. Shang Yang, the radical reformer of the Qin Empire, ended up dying a tragic death; but reformers like him must never be forgotten, nor should their contributions and the lessons that can be drawn from their lives. If he had not implemented the reform so radically and in such a hurry, if he was possessed of a more far-reaching vision and had aimed for gradual reform over centuries, he would have succeeded. Sometimes, the process of reform may last 1,000 years: religious reform in Europe is a good example.

Again, in 2012, at the 100th anniversary of the Xinhai Revolution, Ren Zhengfei read many commemorative essays. He said, "The surrender of the Qing emperor is the greatest compromise in Chinese history. He had avoided war and bloodshed. This reflects the political wisdom of the Chinese people."

Ren Zhengfei regularly draws inspiration from historical movies and programs. For instance, after watching *The Emperor in Han Dynasty*, he started to think about how to handle personal honor and disgrace. Afterwards, he became fascinated by a series of historical lectures aired by Phoenix TV: *Bloody Dusk*. Ren recommended these lectures to other members of senior management. The lesson he gleaned from this series was that leaders must be flexible under changing circumstances: that they must be able to go on the offensive or hold back, insist on principles, or accept failure as circumstances demand. Another program that inspired him was the TV series *Drawing Sword*, which taught him that an outstanding general must be tested on the battlefield and that although young soldiers may have certain shortcomings, leaders must learn to appreciate their merits and tolerate their limitations, as they will be the future generals.

History is an inexhaustible source of ideas. Any history of a nation is a history that reflects its national character, a history of organizational reform that is built on those special characteristics. Countries are the largest form of social organization and they are a sum of many different elements. The rise and fall of a country, including the fate of its leading figures and major events that occur throughout the process, are a massive spiritual legacy for later generations.

It is impossible for a state, enterprise, or individual to break away from history and start all over. But no organization would ever succeed if it only lives in the past, unable to criticize or learn from history. The key is for the organization to open up its mind to the lessons of the past and the outside world.

Political cartoonists have been active in US society for centuries, especially in its political sphere, and they have injected a lot of humor into bipartisan politics. Politicians who are portrayed in caricature form do not feel offended; instead, they are amused or find in these depictions a source of comedic power. These portrayals also help demystify politics and the politicians themselves: Politics is a profession and nothing more, and politicians—well, you know how politicians are; they can sometimes be clowns, too.

The bipartisan politics of the United States has been depicted as a well-known fight between donkeys and elephants. This metaphor originated in cartoons by Thomas Nast, a German political cartoonist who is widely credited with perpetuating the donkey and elephant as symbols of the Democratic and Republican parties. As time went by, both parties embraced their mascots. Even though the donkey may seem dumb to a Republican, Democrats associate the symbol with intelligence and courage. Similarly, the elephant is flashy and cumbersome in the eyes of Democrats, but is a symbol of dignity, power, and wisdom to Republicans.

Since then the race between the donkey and the elephant has become a symbol for the basic trend of rotating power in the United States. The Democratic donkey, representing the middle and lower classes, insists on increasing taxes on the rich, improving social welfare, and reducing the gap between the rich and the poor: it is essentially a wealth-consuming party. The Republican elephant, on the other hand, represents the rich class and calls for tax cuts to boost economic vitality; it is a wealth-creating party. In general, when the Republicans are in power, the United States enjoys an economic boom, but the wealth gap widens. When the Democrats rule, the middle and lower classes gain more benefits and social conflict is diminished, but the economy more often than not enters into recession. This has been the cycle of US politics, economics, and society over the past century.

The clash between the donkey and the elephant, however, has rendered US society as non-donkey and non-elephant. The country has progressed in a balanced way as a result of seeing through each other's motives, and with a sense of agitation from being jostled back and forth between political ideologies. What inspiration can Huawei's corporate culture draw from this kind of national structure?

Huawei hired Mercer, a leading global consulting firm, to design its decision-making processes and mechanisms, which reflect a lot of Western wisdom. Yet the most critical part of Huawei's decision-making hierarchy, the rotating CEO system, was proposed by Huawei executives out of their understanding of Western—or more exactly, US—political tradition. The rotating CEO system is meant to prevent the company from going too far Left or Right and to create a balance of power. The passing years have witnessed an improvement in Huawei's decision making through this borrowed mechanism.

And what has Huawei learned from the constitutional monarchy system in the UK?

Everything is based on pragmatism. Huawei would never label itself as the successor of any single culture. Some argue that Huawei has been able to achieve global success because the company and its senior executives, including Ren Zhengfei, have been Westernized. This is in fact another white-or-black misjudgment. There is not a single totem in the mind map of Ren Zhengfei that is not subject to change and Huawei itself was raised to be strong on a rich variety of grains. It retains any cultural element that proves useful and discards anything that's useless, be it Chinese or foreign, modern or ancient.

For example, the value proposition of customer centricity to which Huawei attaches the greatest importance originally came from Western companies. Ren Zhengfei has written a number of essays urging his colleagues to learn openness, entrepreneurship, and dedication from the American people, especially from those in Silicon Valley. Sixteen years ago, Ren Zhengfei passed through Dubai and was amazed. Later, he wrote an essay calling his people to "build a strong Huawei" with "wisdom borrowed from the whole world." In addition, Ren has recommended the Russian film *The Battle of Moscow,* and *The Long Gray Line* is already a must-read book for all senior management.

Huawei has a university of its own: Huawei University. It is an incubator for leaders and the mixing ground of Eastern and Western cultures. More than 100,000 people have attended Huawei University and taken courses on Huawei's management philosophy, systems, values, and code of conduct. They have also witnessed or taken part in frequent clashes of ideas and culture. The following list of lectures from 2002 to 2010 might more clearly illustrate the origin and evolution of Huawei's culture:

- Protestant Ethics and the Capitalist Spirit
- Returning to the Axial Age
- Useless and Useful: The Wisdom of Lao Zi and Zhuang Zi
- The Book of Changes and Ways of Thinking

- About Nothingness and Abstraction
- The Particularity of the Law of War and the Nature of War
- Guidelines and Strategies for War
- The Origin, Basis, and Development of Christianity
- Modern Western Philosophy
- Re-reading *The Art of War*
- Buddhism in China: Zen
- Comparison of Chinese and Western Cultures
- Interpreting Western Art
- Aesthetics and Sensational Wisdom
- Seeing the World through Paintings
- Appreciating and Criticizing Music
- On Traditional Chinese Medicine
- Fuzziness and Tolerance
- Olympics and Greek Mythology
- Religious Background of the Current International Pattern

Ren Zhengfei calls these courses "eye-openers."

From Thought Clouds to Thought Rain

Normative Power and Power of Mind

Amitai Etzioni, professor of sociology at Columbia University, said in his book *A Comparative Analysis of Complex Organizations*: "Power is characterized by the means to secure compliance. Such means can be natural, but can also be material or symbolic. There are three forms of power: coercive, remunerative, and normative." He believes that coercive power may cause physical or psychological pain, remunerative power is dependent on material satisfaction, and normative power inspires moral involvement and secures compliance on the basis of rules.

Based on Amitai Etzioni's concept of normative power, David M. Lampton, the George and Sadie Hyman professor of China Studies at Johns Hopkins University's Paul H. Nitze School of Advanced International Studies, came up with the concept of "power of mind." This form of power creates and disseminates knowledge and ideas to secure support. In some sense, the power of mind exceeds normative power because it covers a range of factors, including leadership, intellectual resources, innovation, and culture.

All four forms of power coexist in harmony at Huawei, a utilitarian organization. Without the application of coercive power, such as eliminating underperformers, holding people accountable for personal and organizational results, and information-security regulations, the company would fall apart. A company is like a troop of soldiers that would be severely punished for refusing to follow orders to attack or retreat or for spreading defeatism. Like an army, a company is a goal-oriented organization with extensive coercive prohibitions in place to hold the organization together.

Remunerative power is essential for a business organization. Seeking benefits and the idea of earning more pay for more work are part of human nature, which no company can ignore. Building on this aspect of human nature, Huawei has developed and insisted on the core value of inspiring dedication among its employees. Its compensation, bonuses, welfare packages, and its employee shareholding scheme are all remunerative incentives.

Coercive power and remunerative power are quantifiable forms of power, and are based on the black-or-white rules of authoritarian culture. Of course, a company can't operate without rules or authority, but their effects are limited. They won't create any sense of belonging among employees.

In an age when credibility is diminishing, loyalty is losing value, and idols are falling from grace, the commitment of Huawei's employees is unique.

Sure, many people have left, but they have remained emotionally attached to this business organization, just as graduates often miss their old alma mater. For two decades, the inimitable image of Ren Zhengfei, the "Boss," has not at all diminished in the eyes of Huawei people, which truly gives us pause.

Why is this the case? The answer lies in normative power and the power of mind. Ren Zhengfei advocates grayness, which encompasses tolerance, openness, and compromise, and therefore transcends traditional black-or-white ways of thinking. This philosophy has introduced lubrication, flexibility, and warmth to the cold machine of business and, at the same time, satisfies human desire for material gains.

More importantly, Huawei's leadership creates a set of totems at each stage of development, which serve as spiritual banners that surpass anything of a material nature; these banners direct the actions of every member in the organization, causing some of Huawei's employees—and their families, friends, and even clients—to comment that Huawei employees are brainwashed by the company. "Brainwash" isn't really the right word because the company does more than that; perhaps "brainswap" is more appropriate.

Every leader is solitary. So is Ren Zhengfei. Why? Leaders are all solitary thinkers. Milan Kundera once said, "Man thinks and God laughs." But the problem is that in this secular world, God is too far away from us, so every organization, whether it is a nation, a company, a school, or a church, needs a leader who keeps thinking and provides direction.

In this sense, leaders are spiritual laborers, puritan travelers in the realm of thought. I have witnessed the journey of Ren Zhengfei in the world of business philosophy, how an idea is formed, reviewed, developed, and systemized. I understand that this is an incredibly tough journey that requires a special brand of resilience. Ren Zhengfei likes to describe this process with the concept of a cloud. He has said, "It takes more than half a year for a cloud to turn into rain." What he means is that after an idea is conceived, there is still a long way to go before it can be implemented. It has to turn from gray to white, blurry to clear, and relevant rules and mechanisms should be developed to guarantee its effective implementation.

The most solitary period in this process, however, is the formation of a cloud in the sky of thought. Most commonly, Ren Zhengfei gets ideas for Huawei's management and development while reading a book or talking with someone. These ideas then condense in his sky of thought and form points, which gradually connect into lines after more reading and conversations with different people. Ren then speaks about the same subject on various occasions; the lines begin to expand and he proceeds to piece them together into patches. Afterwards, he presents these patches of thought at executive meetings in which they are discussed and debated until a consensus is reached. The final idea is then publicized in a speech or essay that expands in ripples among Huawei's employees, like a stone cast into a pond.

Generally speaking, it takes two or more years for Ren Zhengfei to form a cloud of thought, and another six months for the company to turn that cloud into rain.

Dialectics and Metaphysics

In the Latin quarter of Paris, on the left bank of the Seine River, there are many cafés housed in classical buildings, and people are attracted by the fragrance of coffee that permeates the streets. They walk into a café and spend a cozy afternoon with a cup of coffee and a book.

In such an elegant, quiet, and yet cramped environment, many great European philosophers, artists, and writers have garnered inspiration, including Jean-Paul Sartre, Albert Camus, and Alfred de Musset. Coffee has sparked many great ideas that have ultimately changed the course of human development. However, an article titled "Coffee: An Awkward Plant," published in *Life Week,* reminds us of the other side of the coin:

> It is less known, however, that coffee and cafés have gone through a bloody history. In the 16th century, the darkest age in human civilization, coffee was considered just as wicked as pagans. Conservative theologians in the Arabian Peninsula destroyed all coffee beans on the streets of Mecca, and the prime minister of the Turkish Empire put a café owner into a bag and threw him into the Strait of Bosporus.

Times have changed. Perhaps this is a manifestation of dialectics. Beauty and ugliness, warmth and coldness, justice and evil, success and failure, and right and wrong are all likely to change into each other over time.

Huawei's leadership believes the following:

> Our faith in grayness and compromise is based on dialectics. With an awareness of grayness, we are able to broaden our prospects, see the future more clearly, and hold our course. Our commitment to grayness and compromise does not mean that we are weak; it means we are strong. We may have to plan our strategy over a span of 10 years or more, so it can't always be written out clearly in black and white; there will be small revisions—or even a complete reversion—over the course of time. Such adjustments within a strategic framework are natural and necessary as circumstances change. It's normal to make ongoing adjustments within a broader framework, necessary even. But we can't make any mistakes when it comes to choosing our main direction. Our framework needs to be broader, grayer, and even more blurred around the edges so we won't be too far off in the direction we take.

Ren Zhengfei encourages senior executives of the company to meet with influential people in the world while enjoying a cup of coffee. The idea is that they should increase their contacts with the outside world in order to get more information and develop a longer and broader vision. He said:

> Huawei must stand against dogmatism and the belief that all knowledge is derived from personal experience. If we follow a fixed course, or depend too much on our past experience, we will get lost midway. Similarly, if we rigidly follow a certain theory or dogma, we will stumble and get hurt. There are many MBA graduates who can't run a business. Why? Because they got caught up in dogmatism.

Meanwhile, Ren Zhengfei admitted, "I don't mean that metaphysics is necessarily a bad thing. There are some things that we should hold firmly to, mechanically, as if they're doctrine."

Many recognize the Germans as the most no-nonsense people in the world, as if every cell in their body is infused with the philosophies of metaphysics and mechanical materialism, manifesting itself in every aspect of their daily lives. This is perhaps the reason why Germany has the most advanced and most competitive precision-manufacturing sector in the world and has produced the most scientists, financiers, artists, and philosophers in history. Of course, for this very same reason, Germany was also the most dreaded war machine in Europe at one point in time.

In contrast, British people seem to see farther into the future. They show foresight and a better understanding of grayness in balancing compromise against insistence.

It's sensible to borrow wisdom from the past and from the outside world. Huawei has tried to make the best use of every accomplishment in human civilization, attempting to incorporate them into its own framework.

But the question remains: What should be viewed with grayness and what should remain strictly black or white? Over the past two decades, Huawei has developed its own answer. The company can perceive and embrace shades of gray when it comes to strategy and people. These can be subject to dialectical analysis. Tactics can also be gray to a suitable degree and should be adjusted for different circumstances. However, its core values of customer centricity, dedication, and perseverance can never waver or alter course by even a single degree. They are the metaphysical laws—the Huawei Bible, so to speak—for all of Huawei's 170,000-plus employees. And Huawei's leadership has led to its people in studying this Huawei Bible every day, every month, and every year, until they've internalized its message and its very precepts flow throughout their veins.

Meanwhile, people and business operations are treated differently at Huawei. While people are viewed with grayness, business operations has to be handled with black-and-white process. The processes of product development, sales, delivery, and after-sales service must be implemented without fault in order to fulfill the company's commitment to customer centricity. The reason why people should be viewed with a certain amount of grayness is that people are growing and changeable, and a gray approach is necessary to unleash their drive and creativity.

At Huawei's research institute in India, each Indian engineer writes 2,000 lines of code every month and every Chinese engineer churns out about 20,000 lines. The Chinese employees seem to be 10 times more efficient, but the problem is that each line of code written by the Indian engineers is valid, while for the Chinese engineers only 200 of the 20,000 lines function correctly. It's clear that this problem can't be dealt with using a dialectical, gray approach. In this case, scientific research belongs to the realm of metaphysics, or solid, universal truths.

This contrast has left Ren with a deep impression. He has said that nations with strong religious faith have many merits that are worth examination, such as integrity. He once recommended that Huawei's senior executives and customers read an essay called "Market Economies with Churches and Market Economies without Churches" by Professor Zhao Xiao.

Huawei's senior executives, however, were more impressed with what they saw in front of them: the heartfelt respect that their Indian employees have for intellectual property. For many years, Huawei had a major issue with employees joining up with external forces to steal the company's technology. This is a moral and legal issue, however, with no "gray area" that's up for interpretation; black and white must be firmly kept apart.

Ren Zhengfei said:

> The decision-making process is gray, so decision-makers have to open their minds and learn to compromise. By doing so, they will be able to gather as much wisdom as possible. People at the lower levels of the organization are mainly responsible for implementing decisions. They must be practical and quick-handed. In short, the decision-making process can't be done in haste; leaders need to slow down in order to minimize mistakes, whereas implementation should be as quick and efficient as possible.

Commenting on Ren Zhengfei, rotating CEO Guo Ping said, "For 20 years our boss has seemed very lofty and aloof, but from time to time he'll drill down into the lowest levels of the organization to stir things up at the core. Of course, the organization is able to quickly find its balance again. Huawei has developed a self-healing mechanism."

Self-criticism: Vitality Is the Soul of an Organization

Chapter 6

Black Holes and the Second Law of Thermodynamics

The Theory of Entropy and Organizational Fatigue

In 1854, Rudolf Clausius, a German mathematician and physicist, introduced the concept of entropy. The idea is that, in an isolated system, molecular thermal motion tends to shift from a state of concentrated and orderly alignment to a state of dissipative and chaotic disorder. Throughout this spontaneous process, entropy keeps increasing. When the system reaches maximum entropy, it enters a state of motionlessness, or thermodynamic equilibrium. Entropy is the central idea behind the Second Law of Thermodynamics. This law also implies that the increase of entropy is irreversible. In other words, entropy is an arrow of time, according to Sir Arthur Eddington, a British astronomer. Time is corrosive, leading every creature in this world and the universe to irreversible destruction.

As Karl Marx put it, from the day we're born, human beings are already marching toward their graves. This is a constant law, dubbed "the irreversible increase of entropy" in science.

In 1981, a book called *Entropy: A New World View* hit the United States. This book applied the concept of entropy in natural science to human society. Sixty years prior to the release of this book, English radiochemist and monetary economist Frederick Soddy had asserted that, in the end, the law of entropy would govern the rise and fall of political systems, the freedom or bondage of society, the fortune of commerce and industries, the origin of wealth and poverty, and the general physical welfare of people.

In other words, the emergence of life is for the purposes of survival, not death. Unfortunately, however, all forms of life will eventually come to an end, including animals, plants, humans, nations, armies, companies, mountains, seas, even the whole world, and the universe itself.

The prelude to the disappearance of life is fatigue and aging. One may remain healthy for all one's life, but as time passes, one's genes will decay, one's metabolism will slow down, and one's vitality will decrease. A flower is brilliant in spring when it is blooming, but it will wither away in autumn.

This is the same with social organizations. A dynasty may appear robust and full of life in its early stages as a young and vigorous regime, charging boldly forward as if on the back of an unstoppable steed. Before long, however, laziness, hedonism, and corruption will arise and spread. Fatigue will gradually seep into the bones of the dynasty, even if it suffers from no other social illness (which is hardly possible). And when healthy cells are no longer young, full-on lethargy takes hold.

Business organizations decay at a faster pace. Within years, perhaps no more than a decade, an otherwise robust organization may begin to suffer from laziness and an unfolding cluster of negative energy that grabs at the company like the arms of an octopus. Why? Because pragmatism pervades every single cell of the organization.

Beyond the irreversible passage of time, another reason why organizational fatigue occurs is because organizations tend to depend on past factors of success. After the fall of the Berlin Wall in 1989 and the breakup of the former Soviet Union in 1991, Adam Smith's invisible hand saluted capitalism in its victory the world over. Some free-market fundamentalists in the United States declared, "The world's economy has now entered a period of stability." And as fate would have it, just a few years later, the global economy got caught up in a massively destabilizing whirlwind.

As Japanese sociologist Katsuhito Iwai said, "The enemy of capitalism, or the enemy of freedom, is laissez-faireism rather than socialism."

Companies are no different. Once-blazing-and-influential names like Bell Labs, Motorola, Nortel, NEC, Sony, Nokia, and Alcatel, soon lost their spark, slowly succumbing to all manner of illness and disease. Barring other internal and external factors, there were two important causes: They were too old and they used experience from the past to cope with modern challenges. It's pretty much impossible to adopt such a relaxed and untroubled approach to our rapidly changing world of extremes.

Organizational fatigue has dozens of causes, of which "history" is the most fatal. Ren Zhengfei is afraid of death, aging, and fatigue, which is precisely why you won't find any definitive books or depictions of Huawei's history within the company. Huawei's chronicles spanning 27 years are full of stories fraught with ups and downs, narrow escapes, and crowning achievements side by side with suffering and hardship. Huawei has broken one world record after another, and with so many foreign politicians and business leaders visiting Huawei from all over the world, it really ought to have its own history museum!

In April 2001, while touring the Panasonic Corporation's museum in Japan, I made this very proposition to Ren Zhengfei. He answered, "Huawei does not need a museum. Huawei should forget its history."

In Huawei, the most impressive place is probably its product showroom, which only displays new products and a wall of patents; there is not a trace of the company's history.

History is likely to make people lazy, depressed, disinterested and irritated. German philosopher Arthur Schopenhauer said, "The two foes of human happiness are pain and boredom." In this age of information and globalization, idols, stars, and leaders can only maintain their freshness for a short period of time. Alienation becomes the status quo and loyalty a luxury.

Huawei doesn't celebrate the past glory of its stalwart heroes. Ren Zhengfei won't allow the company to have too much baggage—and that applies to himself, too. As one Huawei executive put it, "The company won't put anyone up on a pedestal, not even the boss. Everyone is viewed in light of their present capabilities and contributions. At Huawei, 'the past is a blank sheet of paper.'"

Black Holes in the Organization: Corruption, Factions, and Indolence

Ren Zhengfei and I both took notice of a news report at almost the same time: Astrophysicists had observed two "rogue black holes," over 10 billion times the size of the sun, wandering the Milky Way 300 million light years away from Earth, threatening to swallow anything that got too close.

According to other reports, NASA discovered an eye in the depths of the universe: NGC 4151, a neighboring spiral galaxy located 43 million light years from Earth, which contains an actively growing supermassive black hole in the middle—a giant white pupil of light. Gizmodo.com called it the "evil eye" of the universe.

Ren Zhengfei saw this news on TV, which didn't go into too much detail, so the next day he went on the Internet to do some further reading. Many years ago, when the president of Peking University visited Huawei, Ren said that he would like to become a student at Peking University after he retires. He would first study mathematics and then thermodynamics to explore the origin of the universe and the Big Bang. The president said he was very much welcome to join them.

Ren Zhengfei often stresses to the executive team that leaders should possess a broad range of vision. They should know a bit of everything, including astronomy and geography, and at the same time, they should be able to adjust their focus as needed. To zoom in or out with ease, one has to have a pivot, a central point around which all thoughts unfold—for example: What is Huawei's black hole?

It's common knowledge that bees are the "matchmakers" among plants. Without bees, the vegetation of the earth, including grains, would be isolated and gradually die out. This would, in turn, lead humans towards hunger and war. Therefore, the fact that bees are dying in great numbers around the world has aroused widespread concern among food and agriculture scientists.

Recent studies show that tachinid flies are a new threat to bees. The flies lay their eggs on the belly of the bees, which then lose their agility and sense of direction. Affected bees fly away from their beehives at midnight and die soon after. This ultimately leads to colony collapse and disorder, with the sudden disappearance of all the adult bees in the hive.

For human beings, cancer is one of the most fatal diseases. Medical studies have shown that cancer cells are a mutation of normal cells and can grow in any part of the body. Cancer cells don't age, but continue to grow and multiply. They feed on the nutrients of other bodily tissues until the body dies.

The hand of God is mysterious indeed! He seems to have created a universe full of striking similarities. Black holes are the destructive cancers of the universe. Bees die from the eggs that flies lay on or within them. Genetic mutation leads to the growth of malignant cancer cells in humans. And plants wither away from the sudden onset of unknown pathogens in their cell membranes.

Then what are the incurable forms of cancer within an organization? They are individual greed and the collective unconscious of corruption. In the Roman Empire, there were more baths than churches. During the last years of each Chinese dynasty, prostitution and drinking were rampant. In any Eastern or Western country that has passed its prime, the upper classes usually succumb to decadence while the lower classes grow selfish and cold. To sum up, unrestrained desire and people's inability to control their own desire are the primary causes of cancer in all organizations. For any organization, corruption, factions, and indolence are the three black holes that accompany it throughout its lifetime.

These days, Wall Street is the most untreatable form of cancer in the global economy. It first struck Western financial capitalism and is now spreading to every other part of the world, including China. The real economy is being drained; wealth flows into the hands of financial and technological oligarchs, leading to class conflict and a get-rich-quick mentality.

In 2011, China became the world's largest consumer of luxury goods. Unfortunately, this is not good news: It indicates that the nation is decaying at shocking speed. The "wealthy class" has only just formed in the last 30 years—strictly speaking, more like 20. In the course of a single generation, most of the first group of people to acquire wealth have already lost their enterprising spirit; they have turned into a class of consumers. In any large or mid-sized city in China, from the very north down to the very south, there are a lot of clubhouses that provide world-class luxury entertainment. Their most frequent customers are the newly rich, who no more than 10 or 20 years ago had been forced by hunger and scarcity to fight, to go on adventures, and to exploit new frontiers.

Ren Zhengfei is wary of the situation. He tells his colleagues and subordinates that they must understand the reasons why past dynasties have fallen. He says that a new emperor would often overthrow his predecessor at a very low cost, because the incredible greed of the royal family in the former dynasty had already driven the country into destitution. However, the emperor of the new dynasty would give birth to dozens of children and each child would become a hedonistic lord dependent on the national treasury. After dozens of generations, a huge parasitic class would form, wreaking havoc on the dynasty. The laboring class, who could no longer bear the burden of the royal family's excess, would rise up and overthrow them. Thus the same old story repeated itself. Huawei would go broke in just a few years if it were to fall prey to similar decadence. Quoting a writer from the Tang Dynasty, Ren reminds his staff: "The people of the Qin Dynasty didn't have time to grieve for themselves, so later generations grieved on their behalf. But if later generations grieve for the Qin, and fail to learn from their mistakes, then the only ones left to grieve for *them* will be the generations after."

Judging Huawei with a Critical Eye

Huawei has remained full of spirit for the past 27 years, which is rare in the business world. Performing one miracle after another, Huawei has been called "a whole other organizational animal" by management scholars around the world. (Ren Zhengfei also likes to compare organizations to animals.) However, success tends to make the successful lose their grip on reality. Pride, lofty aspiration, and desire can lead to an overinflated sense of self, which allows arrogance and the delusion of omnipotence to infiltrate the entire organization.

One day in the beginning of 2013, Xu Lixin, the former vice president of Huawei University, was in Ren Zhengfei's office, where he proposed that a book should be written on Huawei—they could call it *Judging Huawei with a Critical Eye*. Ren Zhengfei applauded this idea: "Excellent! This idea is the best thing that's happened today." Unfortunately, nobody has taken it upon themselves to write this book yet.

I spent one and a half years interviewing over 100 senior executives and experts at Huawei, from whom I gathered some sharp, straightforward, and rational criticism of the company. The following is a summary of what they said:

- Huawei is successful in two aspects. The first is that it has truly made customer needs the number one driver of its business. The idea of customer centricity isn't empty talk at Huawei. Why have we realized wide coverage throughout the entire world? In fact, it's all about getting closer to our customers. When we first started out, why did we set up more than 30 rep offices and 20 customer service centers in China, and why have we established our presence in 170 countries and regions? It's because we want to be where our customers are. The voice of our customers is the highest form of command in the company. In the past, we won customer respect not because of our technology, but because we have a sort of self-healing mechanism: whenever and wherever there was a problem on the customers' side, everyone in the company was willing to jump on the problem and find a solution. And whenever an internal problem occurred, no one would shirk responsibility or start pointing fingers. Things are different these days. People have started placing blame on one another. Take the conflict between the frontline and the headquarters, for example. The frontline communicates with customers to understand their general needs— they look at the "general ledger" so to speak. Headquarters, however, looks at the "itemized account" and measures the performance of the frontline based on each item. This has been the cause of endless dispute, because evaluation has become too granular. Is it really necessary to measure performance at the individual product level? This reflects the immaturity of our management. Managers pass down market pressure level by level, and the people below them start assigning tasks and responsibilities to different teams. This approach is intended to hold each product line accountable for its business operations and results, but in reality it's made people a little crafty: They've invented quite a few clever—if not dishonest—techniques to make sure they get the right results. It seems as if everyone is focusing their effort on internal conflict, not on satisfying customer needs.

- Huawei started out selling switches, but we rarely told customers that we were selling switches. We sold campus cards, business networks, and so on. There are an increasing number of technical terms these days, and hardly any of them are customer-friendly: 5G, 400Gbit/s, VDSL, etc. Can you catch any non-technical words among those? These days we don't speak the way our customers do, like we did in the past with words like "distributed base station" or "SingleRAN." This is an extremely significant trend that goes against Huawei's value of customer centricity.

- In the 1990s, Huawei was the epitome of advanced productivity. Are we still that way today? Think about Microsoft, which is making great profit, but whose era has already passed. What about Huawei?

- What's it really like in Huawei today? Many people in the company don't listen to customers, but to their bosses. In the early days, no one in Huawei was afraid of failure or responsibility. But today people are afraid of making mistakes. Senior executives are exhausted at work, running about like crazy from one meeting to another. The middle management team does not make decisions, because they have to be responsible for the decisions they make. The top management team goes on about getting rid of "crutches," but in reality crutches are built on top of other crutches. Too many departments aren't actively engaged in the business, and we seem to only hire a bunch of automatons. Since when did we stop hiring heroes? In the past we brought in all sorts of people, and they were all able to find a way to stir things up a bit. Now everyone's so deferential and obedient. Now that we've got the Internet, many managers don't read any more. They would rather look at a bunch of "chicken soup" stuff on social media. They talk about buying houses and cars all day long, but nobody talks about how many books they buy in a year. If they don't read and they don't think, where does Huawei's future lie?

- A major issue at Huawei today is that people don't stay in tech-related roles long enough. If you're still doing development work in your thirties, people are going to look down on you—you'll look down on yourself, too. If you're not a manager or something, you'll either be shoved off in the corner somewhere or kicked out entirely. This type of mindset discourages people from settling down to develop technology. In other countries a lot of people are still engaged in R&D work well into their thirties and forties. They might not have the same amount of energy, but they have rich programming experience that helps with efficiency and productivity. Whereas over here, we're turning a bunch of sharp technology and domain talent into managers.

- Among the 170,000 employees in the company, how many are truly making themselves useful? At most 70–80 percent, and that's optimistic. The rest are idling the time away. And it's not because they want to. Do employees lack money? No, they lack opportunities. Unfortunately, the company doesn't create opportunities for them. Twenty years ago when we first joined Huawei, one person was responsible for developing several boards on their own, but now you've got a dozen or two dozen people working on a single board together. In the past, one person would develop the entire mainframe program, but now hundreds of people develop one program. Ericsson's sales revenue is similar to Huawei's, but it only employs 70,000 people—their strategy is to use experts for precision work. We have double the number of employees, and our strategy is to throw them at projects like a tsunami. This is going to be a big problem for the future: You have to increase salaries, otherwise you won't be able to retain competitive talent. But you've got more employees and worse management—how in the world are you going to survive?

- There are always two sides to the same coin, so there can be different options for different periods of time. Huawei isn't as dynamic and aggressive as it used to be. Therefore, we should take a more flexible and pragmatic approach towards organizational design. Take our IPD transformation, for example. It introduced the IPD process, along with relevant organizational changes, which were designed to ensure that we can consistently launch new products to the market and that those products will be successful. This process has been running for more than 10 years, but have we adapted it at all? Is there anything worth optimizing? Can it contribute to the success of our consumer products? If we rigidly stick to the status quo, the vitality of our organization will slowly dwindle away. A process itself is not a living thing, but managers can inject life into it and make sure it keeps ahead of the times. Take our "5/8 principle" as another example. According to the design of our organization, you can't form a department or team without at least eight subordinates; and a department can't belong to an upper tier if it doesn't have at least five lower-level departments reporting to it. I understand that with a company this size, you need clear management planning. But aren't some of these rules a little too absolute? Sure, if you don't want to be a manager, you can still be a technology expert, but only managers have the right to evaluate team-member performance and plan budgets. For a big company, a rigorous, objective, and standardized management system is an important guarantee of sound operations. It's certainly not my wish that Huawei return to the chaos of its past, but let's not forget that rigorous and objective management has to be balanced with a more practical human element. For example,

Huawei employees have to badge in and out each day so the company can track their attendance. I don't agree with this method. Managers should be responsible for managing their subordinates; whether or not you badge in on time isn't directly related with job performance. Managers should carry out their own responsibilities, focusing on the work their people do, not on whether or not they're sitting in their seats. Chaining people to their desks offers no guarantee of productive output.

- Why does headquarters keep on growing? The main reason is that power is not properly delegated. Our leaders enjoy concentration of power and tend to interfere with things that are outside their purview. A single director might need four different crutches to prop himself up, so it's no wonder headquarters has gotten big. As long as they've got power, HQ is never going to shrink. And if you take their power away, they have no reason to exist anymore. Since the power is still there, when someone above you says it's time to streamline, they're basically maneuvering against you. Every department goes head to head with the company. If a department relinquishes power, they're essentially engaging in a revolution against themselves—they're giving up their meal ticket, and no one's willing to do that. Since what you do is linked to your performance, and your performance is linked to your benefits, nobody wants to relinquish any power. Our company's management philosophy is that each department should take care of its own business. You can give up some of your own power, but that only results in a higher concentration of power elsewhere.

- Our boss has said many times that there's no need to spend time washing coal because coal is meant for burning. What's important is that you create value. It doesn't matter who you are or how you do it, whoever creates value for customers is entitled to better pay and job promotion. However, the reality is that in recent years, eccentric or unorthodox people have had a hard time at Huawei. The company's process is extremely complicated and no one has any independent power. Within a network-like system of checks and balances, even if your efforts are recognized by someone higher up in the hierarchy, you might not be accepted by your peers, so those who go against the grain tend to be elbowed out by their colleagues.

- Huawei has to avoid doing things the "Chinese way:" worshipping authority, heroes, and dignitaries rather than relying on systems. If Huawei wants to build an everlasting business, it has to rely on systems instead of individuals or heroes.

- Huawei's culture is like a band of brothers. With friendships that were forged together in the heat of battle, whenever you fight, you stand back up and you're still brothers. If you disagree, you fight it out behind closed doors. If you don't come to an agreement, then you can't leave the room. As soon as you reach a conclusion, you step out and maintain a united front: It's a collective decision that everyone has to implement. It's a very united, combative approach. However, those who join us from the outside have trouble surviving in this type of culture. They don't know how to interact with a bunch of Huawei guys who fought together in the same trench back in the day. (For example, Huawei poached Colin Giles, a marketing genius from Nokia, at a high price, but after working for less than half a year in the Huawei device team, he resigned.) This is a fatal trap for a global company.

In October 2013, Huawei's executive committee of the board issued a notice to the entire management team on an in-depth study of criticism directed at Huawei aimed at driving improvement within the company. The introduction of the notice went like this:

> We have invited our customers to share their impressions of Huawei directly with the Board of Directors. They have been lenient, firing on us with mercy, but we were nevertheless subject to heavy, harsh bombardment. If Huawei does not remain alert, we may one day perish under the gunfire.

After the notice was released, Huawei's managers responded enthusiastically by submitting their own thoughts and opinions. More than 100,000 employees also actively participated in the discussion on Huawei's online forum. Some of the criticism was quite penetrating:

> Our KPI assessments are still too focused on ourselves, with only a little weight given to customer satisfaction.

> It's not uncommon to see firefighter-like heroes gain instant fame in the company, but sometimes these so-called heroes are also the arsonists.

> If people aren't paying attention to product quality, then who will truly care about the quality of our products?! As long as people can close a deal, they don't really care about customer needs. As long as things are delivered on time, they don't care about potential risk. As long as they can keep their positions, they don't care if they cheat their bosses and lie to their subordinates. Our customers are like water, and we're the boat: Water can keep the boat afloat, or it can also overturn it!

> If Huawei was to fall, the primary villain would be those short-sighted performance targets that have infiltrated the whole company. Department silos, pleasing the boss rather than the customer, and a general attitude of disinterest…

> If you simply look at each problem on its own, none of them is a big deal, but a series of small problems will lead to the outbreak of a major crisis. Just as the story goes, "the kingdom was lost, all for the want of a horseshoe nail."[1]

Obviously, while outside praise and recognition of Huawei is on the rise, Huawei's own employees don't think the company is perfect. Rather, they find it full of weaknesses and problems. It's because of this, perhaps, that Huawei still has hope.

All is bright when you're sitting in the dark. These days, what Huawei needs most is to be judged with a critical eye.

---

[1] This is a reference to a famous proverb, which has taken many forms throughout the centuries: "For want of a nail, the shoe was lost/For want of a shoe, the horse was lost/For want of a horse, the battle was lost/For the failure of battle, the kingdom was lost—/All for the want of a horseshoe nail."

Dissipative Structure and Self-criticism

Dissipative Structure and Basic Values

In 1969, Ilya Prigogine, a Belgian physical chemist, proposed his theory of dissipative structures on the basis of the Second Law of Thermodynamics. A dissipative structure is an open system that exists far from thermodynamic equilibrium. In a dissipative structure, through the process of exchanging matter and energy with the outside environment, when heat dissipation and the non-linear dynamics of internal mechanisms cause energy in the system to reach a certain level, the flow of entropy can go negative and the total entropy change of the system will be less than zero. Through a reduction in entropy, the system forms into a new, more ordered structure. According to the theory, more orderly dissipative structures would emerge as various systems evolve or degenerate, driving the physical world into a state of growing diversity and complexity. The theory of heat death held by some pessimists was, therefore, shattered.

This was an incredible discovery. In 1977, as the founder of modern thermodynamics, Ilya Prigogine was awarded the Nobel Prize in chemistry for his contributions to non-equilibrium thermodynamics, particularly his theories of dissipative structures. In a way, this also demonstrated man's optimism about nature and the law of life. On the other hand, Rudolf Clausius, who discovered the Second Law of Thermodynamics, received little recognition among scientists; on the contrary, he had been harshly criticized for proposing the concept of heat death of the universe. Approximately 100 years after Clausius' discovery, although the concept of entropy and the principle of entropy increase were recognized as the most important scientific concept and the most fundamental principle—applicable to human society as well as nature—the theory of entropy itself remained a cold and relentless scientific law.

In contrast, the theory of dissipative structures is much warmer, because it gives human beings courage and hope. It states that so long as we are in an open system, where energy is exchanged and released, any natural or human organization may escape chaos and return to order. Therefore, entropy may decrease and the body may be renewed. This is no doubt the best endorsement of human struggle and endeavor.

In July 2012, Ren Zhengfei said in a speech,

> Later, my interest shifted from heavenly bodies to social dynamics. I care more about how to keep people going and motivating them to create wealth. This is the source of our reform initiatives. This is also the basis of our core values of customer centricity, dedication, and perseverance.

On the topic of dissipative structures, Ren continued:

> If you jog every day, you are in a dissipative structure. Why? You are burning excess levels of energy, turning it into stronger muscle and improving blood flow. As you dissipate this energy, conditions like diabetes might go away, obesity too, and you'll become healthy and beautiful. This is the most basic concept of a dissipative structure. Then there is another question: Why do we need dissipative structures? We say that we're committed to the company, but most often it's because the company is paying too much, which is not sustainable. Therefore, we must first dissipate this type of heated commitment, and substitute it with dedication, with process optimization. There is a difference between working hard and then reaping benefits, and getting handed something before you decide to commit. We must dissipate our latent energy to form new potential energy.

Huawei is determined not only to dissipate the devotion of its people, but to also dissipate their pride. Ren Zhengfei said, "We must dissipate blind pride. What are we proud of, anyway? Instead, we should have a stronger feeling of crisis. We know more than anyone how hard our lives have been." He further stated:

> We have been alternating from stability to instability, from equilibrium to non-equilibrium, from certainty to uncertainty. We are doing this over and over again to maintain the company's vitality. If you eat a lot of beef and don't exercise, you will become fat. But if you exercise hard enough after eating beef, you will

become a tough athlete. In both cases you are eating beef, but you end up in a completely different shape, depending on whether you dissipate energy or not. Therefore, we are determined to hold on to this system.

The system he's referring to is a dissipative system. What would Huawei dissipate? Rich man's disease, laziness, and hedonism. And then, hard work and perseverance could pack some muscle around Huawei's culture. Dedication is the cornerstone of Huawei's system of core values, and the primitive gene that has enabled Huawei to grow and mature. As the company flourishes, however, and conditions continue to improve, this primitive gene might mutate. In fact, some people and some departments have already begun to change shape. Because of this, in 2006, Ren Zhengfei mentioned that "when we talk about dedication, it's not only physical hard work, but more of a mental dedication."

It's easier said than done, however. A perfect idea might deform and decay when it materializes in strategy and tactics, its value reduced to an empty slogan and finally forgotten. Then how has Huawei managed to effectively instill its values throughout the company? While it has surely helped that these values have been cited and repeated almost every day of every month for the past 20 years, the company's insistence on self-criticism is all the more essential.

Self-criticism is an important tool to enhance and firmly establish core values. In a turbulent age when people are torn apart between rapid economic growth and ethical decline, companies must try to preserve the uniqueness and purity of their own values. But this, of course, is an extremely tough challenge. Self-criticism is no doubt the best, least-intrusive means of suppressing entropy within an organization.

However, when self-criticism isn't able to hold back the irreversible increase of entropy, an organization needs to introduce negative entropy flow through radical reform to offset that increase. In this way, the organization will be able to mend its disorderly state and regain vitality. Huawei has conducted a number of radical reforms, including process transformation, mass resignation of the sales department, and the collective buyout of 7,000 people's seniority within the organization, which have reshaped the organization, dissipated subgroup politics, and broken down hierarchies, placing people with a stronger sense of mission, crisis, and hunger at the forefront of the company.

Self-criticism and organizational reform help prevent situations where, because of cronyism or interpersonal politics, the bad eliminate the good—or "reverse dissipation"—from occurring at Huawei. A healthy and growing organization must first of all possess a strong ability to heal and restore itself, which means its healthy cells should be able to eliminate and replace decadent and degenerated cells, not vice versa.

Going against Instinct: The Core Proposition in Organizational Management

Like a person, an organization in its middle age often feels tired and gives off an air of exhaustion and rancid decadence. It's only natural. Organizational fatigue and corruption, as well as the greed of its individual members, may then give rise to degenerative subcultures, such as subgroup politics, leader worship, collective inaction, treason, and mutiny.

Therefore, leaders of organizations spend their lives fighting against organizational fatigue and corruption. The Roman emperors, the first emperor of the Qin Dynasty, Vladimir Lenin, George Washington, Mao Zedong, Margaret Thatcher, Bill Gates, Louis Gerstner, Vladimir Putin, Jack Welch, Silvia Berlusconi, and Adolf Hitler have all experienced and witnessed the madness and weariness of their followers as well as their own. These dramatic changes are due in part to the imperfections and faults of the leaders, but the real killer is the ennui caused by the passage of time.

In history, bloody revolutions and blood-changing reforms have all been intended to punish decadent organizations and conquer the fatigue of nations. The multipartisan political system of Western countries and the leadership transition system of China, from the perspective of organizational psychology, are both based on profound insight into human nature: Everyone expects new faces, which will bring fresh enthusiasm.

In human history, great thinkers such as Sakyamuni the Buddha, Jesus Christ, Confucius, Socrates, Karl Marx, Sigmund Freud, Max Webber, and

Peter Drucker were organizational physicians who had tried to examine the health of social organizations (including states and companies), diagnose them, and prescribe remedies for their various illnesses. They surveyed the times, systems, societies, and countries that surrounded them with exceptional coolness and detachment, but in the end, they gave encouragement and warmth to the masses.

Both organizations and individuals are instinctually drawn towards an increase in entropy, which is a form of degeneration that occurs with the passage of time. However, humans are not helplessly controlled by their instincts. A counter-instinctive approach is a core question in organizational management. Ren Zhengfei once said, "We must constantly engage in self-criticism. No matter how much progress we've made, we still need self-criticism. The world has developed through eternal denial of itself." Why does Ren Zhengfei repeatedly emphasize self-criticism in Huawei? In fact, this reflects his profound insight into humanity and the nature of an organization.

The purpose of self-criticism is to keep enhancing each member's ability to endure discomfort within the organization. The instinctual siren call of comfort affects an organization's behavioral patterns, the performance of individuals within the organization, its reaction to adversity, its decision making, and its process of aging. Huawei's long-term commitment to growth through self-criticism aims to integrate the "uncomfortable" state or the culture of dedication into the organization's DNA and at the same time guard, resist, control, and change the "free fall" of the organization in order to keep the company permanently out of the its psychological comfort zone.

For great thinkers, criticism is a sharp scalpel, and the reforms that result from their criticism have steered organizations away from fatigue and disease while instilling them with newfound life. In the 1840s, Karl Marx, a Western intellectual, published a book, *The Communist Manifesto*, which declared the impending death of capitalism with a sharp voice like an owl screeching in the dark. He argued that the binary opposition between the bourgeoisie and the proletariat would lead to the ruin of capitalism.

Karl Marx, it turns out, did not kill capitalism; instead, he helped pave the way for it. His resounding "death knell" inspired another declaration about 100 years later: *The Capitalist Manifesto—How to Turn 80 Million Workers into Capitalists on Borrowed Money*. The idea was that people could gain income through both labor and capital investment. The book was coauthored by Louis O. Kelso, who is chiefly remembered today as the inventor and pioneer of the employee stock ownership plan (ESOP). The man behind the rapid growth of the middle class in the United States since the 1960s, Kelso has helped prop up the roof of the capitalist framework.

In 1988, 10 years after China began reforming and opening up, a publisher friend of mine said with a hint of dark humor, "Karl Marx's *Das Kapital: Kritik der politischen Ökonomie* (Capital: Critique of Political Economy) has been widely read in China, but it seems like the real 'capital' was left behind in the West, while only 'political critique' was given to the East." In 1992, during his tour to the south, Deng Xiaoping said, "'Do Not Debate' is one of my greatest inventions." For about 30 years, following the principle of "Do Not Debate," China has created the most remarkable golden age in its history, and at a miraculous speed. Of course, in this process, China has also acquired its fair share of national fatigue and illness.

Today, China seems to be offering more freedom and room for debate while keeping its hold on capital too. Finding the right balance between the different, often conflicting dimensions of capital and political critique will test the wisdom of the entire nation—particularly for the social elite, including entrepreneurs.

Escaping the Cycle of Organizational Fatigue

In 1948, Huang Yanpei, prominent educator, industrialist, and politician, who was also one of the founders of the China Democratic League, visited Yan'an, the wartime capital of the CPC. He said to Mao Zedong with almost abrupt frankness:

> For 60 years I have heard with my own ears and seen with my own eyes the rapid rise and fall of individuals, families, groups, and regions. Few have been able to avoid this swift cycle. In most cases this is because, at the start, one is focused and dedicated despite difficult conditions. People are forced to fight for life, or

for a better life. As the environment gradually improves around them, they gradually lose their will along with it. In some cases, laziness arises as a natural result of fatigue, and the feeling spreads wider until the whole organization gets infected. At this point, no one can turn the tables around. Among the people that I know in the Communist Party, most have been fighting for a new path, one that will help us escape this deadly cycle.

Mao Zedong answered,

I believe we have found a new path, and that we can escape this cycle. The path is democracy. If the people are allowed to oversee and criticize the government, the government won't allow itself to grow slack. If every one of us is responsible for this, what we've built won't collapse when we are gone.

Mao Zedong was a great man who slept with his eyes open. Throughout his life, he had sought to understand the natural law behind the rise and fall of states and political parties. In order to prevent organizational disease and national fatigue, he came up with a unique invention: movements. Every 3–5 years, he would initiate a movement to stir up the entire country, involving almost every person at every level of the CPC and China. These movements attempted to awaken individuals and organizations that might have otherwise fallen to laziness and degenerative illness. In essence, the lifelong pursuit of this poetic politician was his ideal of achieving national stability by creating uncertainty throughout the country.

"The people of China face their greatest danger"—this is a line from the national anthem of the People's Republic of China and has served as a warning call for the whole nation since 1949. During the past years of ups and downs, the CPC, the ruling party, has insisted on the principle of "criticism and self-criticism" and has applied it to all people and organizations throughout all periods of development. This alertness to crisis has served the country well.

As a member of the CPC, Ren Zhengfei has experienced almost every change in China over the past 60 years. He also served in the military for several years. Therefore, in his business management, it is only natural that he has been deeply influenced by the CPC governance and culture, just like most of his contemporaries.

Worth noting is that Huawei's concept of self-criticism is a revision of the CPC's "criticism and self-criticism." Criticism means to create alertness against danger. In the history of China and the world, many extremely powerful organizations collapsed easily when disaster struck. The great Qin Dynasty of China that conquered and united the whole country lasted only 14 years. It didn't face any hugely powerful enemy; rather, the dynasty was killed off by a poorly armed peasant uprising. Again, throughout 30 years of reform and opening up, China has produced a large number of star companies and business people, but how many of them are still on the stage? In most cases, they died off the minute they gained a name for themselves,[2] because they lost their sense of fear and caution and fell short on self-criticism and introspection.

Ren Zhengfei often warns his senior executives that wood turns into ash after it burns bright. He requires them to read more about Chinese and foreign history and recommends books, essays, and TV programs that he himself found edifying. He asks them to pay special attention to tragedies that have ruined nations and companies due to internal corruption. He has said, "The biggest risk facing the company will come from inside the organization. Huawei has upheld the tradition of self-criticism; that's our secret weapon for conquering this internal crisis."

Maurice Greenberg, founder and CEO of the American International Group (AIG), is a legendary figure. He is short in stature, yet a financial giant. He once asked Ren Zhengfei, "What is the key to Huawei's success?" Ren Zhengfei answered, "Self-criticism. Huawei has advanced forward through the negation of self."

Greenberg said, "It's a shame we didn't meet sooner. If we had met 30 years ago, I would love to have invited you to work with me."

In fact, the United States, which is the top superpower in the world, has reached this position because it has maintained a sense of crisis for

---

[2] Translator's note: The author employs a fun phrase here, which literally means "like a crab, the minute they turn red, they're dead." The color red in Chinese is often used to refer to fame and fortune.

centuries and because it values the tradition of self-criticism. Of course, there has always been a slew of critical voices in the country: None of the political parties, politicians, social elites, or presidents have ever let down their guard for a second. Today, however, perhaps because it has been the world leader for too long, politicians and business organizations in the United States have begun to grow tired, restless, and arrogant.

Huawei's leadership hasn't just spent its time copying ideas from books or other organizations. Huawei has derived its corporate values of "more work for more pay" and its organizational tradition of self-criticism from the CPC. Ren Zhengfei said, "We have learned all this from the CPC." Unlike the CPC, however, Huawei does not advocate criticism between colleagues. Ren Zhengfei explained:

> Mutual criticism is likely to hurt one another and create tension in the company. Self-criticism is much better. Everyone knows their own boundaries, and can criticize themselves to that extent. While those who dare to reveal their weaknesses and are willing to get help from others will progress more quickly, people who are afraid to lose face will become thick-skinned over time and eventually catch up with the growth of the organization.

Self-criticism: Direction, Approach, and Rhythm

For over 20 years, Ren Zhengfei has spoken most often about Huawei's core values and its tradition of self-criticism. The core values are the company's spiritual totems, ensuring ongoing victory in the midst of fierce market competition. And self-criticism is the shield that protects the integrity of its core values. Both are equally important to the company.

When Huawei's Charter was finalized in 1998, Ren Zhengfei proposed that they place a stone marker at the front gate of the company's new campus with the inscription: "A company's long-term success depends on the continuity of its core values and its tradition of self-criticism."

Some executives at Huawei pointed out that it's wrong to portray Huawei or Ren Zhengfei as too perfect. In the course of its development, Huawei has done as many right things as wrong things, and many mistakes were both foolish and immature. Furthermore, a boss is a boss, not a god. In 2000, when the company held a closed-door meeting to discuss the future, Ren Zhengfei talked about how he could buy three Mercedes-Benzes, one for regular use, one to occasionally drive around for a change, and one to park in the corner, say, to serve as a chicken coop. Consumer demand for physical objects would ever grow as people could always find a use for them. But the same can't be said for information. Information takes time to consume, so how could he possibly need that much information? At the time, he believed that the telecom industry had already reached a ceiling in terms of information saturation. Finally, another perspective gained the upper hand in the company: what they call "corner overtaking," or sharp maneuvers at critical moments of industry change in order to drive ahead of the competition. "The argument was that, if Huawei couldn't do certain things, it's likely that its competitors couldn't either. As Huawei had less baggage, perhaps it would have more opportunities too."

Ren Zhengfei himself admits that he's nothing more than a village-raised intellectual—20 years ago, he managed to start his business not because he had some grand vision for the company's development in China or the whole world, but simply because he had the courage to engage in trial and error and to start again from scratch when he made mistakes.

Many senior executives at Huawei share the same opinion: Without repeated trial, error, and correction of mistakes, Huawei would not be what it is today. Over the past 20 years, Huawei has stuck to one premise—non-stop self-criticism based on customers' needs. "Our boss talks about self-criticism as frequently as he emphasizes the company's core values."

Let us see how Ren Zhengfei interprets self-criticism. In 1998, he said, "Whether or not Huawei collapses depends entirely on itself, on whether management can improve. And this improvement depends entirely on whether or not we can make managers accept our core values, and on whether we continue to practice self-criticism."

In a 2004 essay entitled "From the Realm of Necessity to the Realm of Freedom," Ren Zhengfei asserted,

> Only companies that are capable of self-criticism can survive. In this sense, Andrew Grove, former Intel CEO, was only partially

right when he said that only the paranoid could survive. I would add one more sentence to that statement: You have to understand different shades of gray and practice self-criticism in order to survive.

This was the first time that Ren Zhengfei placed self-criticism and grayness in parallel as the two key factors to a company's survival.

In 2008, he gave a speech, "Those Who Can Climb out of the Mud Pit are Saints," in which he systematically elaborated on the importance of self-criticism in Huawei's history:

> The past 20 years have shown us how important self-criticism is for a company. If we had not insisted on self-criticism, Huawei could not have come this far. Without self-criticism, we would not have listened so attentively to our customers' needs, nor would we have looked to learn from the merits of our peers. In short, we would have fallen prey to egocentrism, and in consequence would have fallen out of this rapidly changing and competitive market race. Without self-criticism, we would have succumbed to any crisis that came our way, because we could not have urged ourselves onward, using the remaining embers of our lives to ignite the fighting spirit of our teams and to light their path forward.
>
> Without self-criticism, we would have closed ourselves up and missed out on the new ideas that have proven essential for us to become a world-class corporation. Without self-criticism, we would have been blinded by pride in our achievements and fallen headfirst into the mud-filled pits that line our path. Without self-criticism, we would not have been able to eliminate the inefficiencies in our organization and processes to reduce the cost of operations. Without self-criticism, our managers would not have told the truth, listened to the voice of criticism; they would not have learned or improved, and as a result, they would not have been able to make or implement the best decisions.
>
> Only those people who have insisted on self-criticism have the right state of mind to accept new ideas, and only those companies

that have insisted on self-criticism will enjoy bright prospects. Self-criticism has led us this far, and will continue to lead us farther. Our future depends on how long we uphold the tradition of self-criticism.

In order to institutionalize self-criticism and avoid straying from its intended purpose, Huawei established a Self-criticism Steering Committee in 2005. At a meeting of the committee, Ren Zhengfei proposed that self-criticism become a key standard for manager evaluation and promotion. He said, "Huawei should change itself from head to toe. We must employ and promote managers who dare to speak up, criticize themselves, and accept the criticism of others. Only such people can fulfill their duties as managers."

In fact, he had expressed a similar view in a speech he gave in 1998 on institutional reform and manager development:

> Any manager who is not capable of self-criticism should not be promoted. We should carefully examine those who have received no criticism. On the other hand, those who have been subject to harsh criticisms should be classified and treated differently. If they have no ethical problems, we should still give them opportunities. People have a natural inclination towards laziness; they are not born with innovative capability. If our managers cannot criticize themselves from time to time, Huawei would soon fade away. In a few years, we will make it clearer that anyone incapable of self-criticism will not be considered for a managerial position at Huawei.

It was made clear enough in 2006. Huawei had spent eight years in preparation for an upcoming transformation program within the organization. This reflects the institutional and cultural character of Huawei: no large and sudden movements, no group movements—and no "movements" of the group. As the ancient Chinese philosopher Laozi said, "Govern a great nation as you would cook a small fish; do not overdo it." This is also true with running a company: A delicious dish may be made only through slow and careful preparation.

Most other Chinese companies, including SOEs, are different. Many are afraid of self-criticism. On the surface, their management seems very harmonious, but at a deeper level, they are at odds with each other. In the long run, these types of organizations—and almost every member in them—suffer from an organizational form of hardened arteries, in which impurities in the veins and arteries of the organization pile up and gradually form into plaque. Some other companies adopt self-criticism too rashly, which can cause hostility within the managing board and ultimately lead the organization to disintegrate. Of either case, there have been plenty of examples among private companies: We have seen slow, suicidal decline for lack of a healthy climate of criticism and seen an avalanche-like collapse from brutal internal struggle.

Huawei's self-criticism has a very clear direction: to revolve around the core values of customer centricity, dedication, and perseverance. Any person or department that goes against these three core values will become the target of self-criticism.

Forms of self-criticism at Huawei include manager feedback meetings, self-discipline declarations, rectification assemblies, and so on, which are all practices gleaned from the Chinese Communist Party. They reflect the Eastern element of Huawei's culture.

Huawei advocates self-criticism, but tries to avoid mutual criticism in its corporate management. Ren Zhengfei believes that a company is naturally a constructive profit-seeking organization, unlike any political organization or government department. Companies are different from military outfits and artist colonies as well. Huawei's approach to self-criticism has a few rules. Its not a test, nor is it used to mobilize the masses. It's focused on oneself, not used to point the finger at others. And finally, it's a fact-based practice, moderate in nature without going overboard. Ren clearly stated that

> self-criticism is not just criticism for the sake of criticism, nor is it intended to deny or negate anything. Our culture of self-criticism should be constructive. Through self-criticism, we intend to optimize and further develop ourselves. Our end goal is to improve the core competitiveness of the company.

This is the underlying purpose behind Huawei's long-term adherence to the practice of self-criticism.

Mental Criticism and Executive Declarations

The Fusion of the Chinese and Western Culture: Tai Chi and Bullfighting

Niall Ferguson says in his book *Civilization: The West and The Rest* that the West has risen to global dominance because it has developed six "killer applications" that the rest of the world lacks: competition, science, democracy, medicine, consumerism, and work ethic. More importantly, it has fostered a culture of self-criticism, which Ferguson calls self-flagellation: "The West is able to find the best cure for any disease. [...] In fact, if any feature is written in the genes of the post-feudalist West, that feature is public participation and accountability."

Huawei is competing with Western companies against a backdrop of globalization, and the battlefield covers all continents, including the home countries of its Western competitors, mainly the European Union and the United States. The competition was asymmetric in its early days, back when Huawei was far behind. Yet with the power of its culture, the enterprising spirit of its people, and a strategy of relentless focus, Huawei launched a fierce race with its Western competitors that lasted for almost two decades. Around 2005, Huawei began running shoulder to shoulder with its competitors, even surpassing them from time to time.

Being on par with each other in terms of products and technology, the competition hinges upon crisis awareness and self-criticism. In other words, in order to beat its Western competitors, Huawei should strive to be more Western than the West and at the same time take advantage of its inherent Eastern way of thinking. In a manner of speaking, the company's ability to occupy the gray area where the skills of bullfighting and tai chi overlap will decide the outcome of this game.

The only way to beat your opponent is to be more like your opponent than your opponent, while possessing special genes that your opponent can't copy. Huawei's culture of self-criticism has left a strong impression on people in the West, and their comments have been very positive. Although Western civilization has grown on the basis of continuous self-questioning and self-criticism, the belief is not that criticism is done for criticism's sake; the aim is to improve and optimize the current system. Huawei is following the same path.

In 1997, the company's 10th-anniversary year, Huawei was still in a state of chaos, though it remained dynamic and powerful. At about this time, Ren Zhengfei started to advocate self-criticism more often, and the company's meetings of self-criticism and peer critique, the so-called "democratic meetings" that had been going on for the past 10 years, were institutionalized and extended to every level and office in the organization. This is a typical CPC practice for organizational development and has helped Huawei in developing its own managers and teams.

All this, however, is more of a call to vigilance, a way to keep people on their feet. It wasn't able to resolve fundamental problems with the system. Cliques, hedonism, nepotism, and arrogance had lingered and even intensified in the company.

In short, the 10-year-old company faced two major challenges: one was conflicting thoughts and goals, and the other was the corruption that had accompanied its rapid growth. Ren Zhengfei's solution to the first challenge was of typically Chinese style: He would propose a general guideline or a master concept to guide the minds and hearts of the people. This guideline was the Huawei Charter, which has played a crucial role in the history of the company. The charter, drafted by professors from China's Renmin University, took two and a half years to finalize. The thoughts and assertions of everyone within the company were thoroughly expressed, argued, and debated. And in the end, it resolved all disputes as to the company's future and direction.

The second challenge was apparently more difficult. All countries and organizations in the East and West alike have had to cope with greed and corruption throughout history. All have accumulated a vast amount of knowledge and experience, but no definitive and effective solution has been found. Is it because the challenge is too tough or too complicated? Perhaps it's because the challenge has evolved over time.

Also, in 1998, Huawei hired a foreign company to design and implement a professional system of management procedures. The idea was to change the company's approach to product development from being technology-oriented to being customer-oriented. For this purpose, the company established a Reengineering Steering Committee and established the principle of "copying, optimizing, and then institutionalizing." For more than a decade, this principle has worked quite well, and Huawei is still optimizing its processes. This Western-style reform aims to build its global customer-service capabilities and make Huawei more like its opponents. This reform has another function: to cage the power of managers, so to speak. Of course, when the company shifts from being governed by people to being governed by processes, people with vested interests will be affected and certainly resist. In response, Ren Zhengfei insisted that "the feet must be cut to fit the shoes."

During the Spring Festival of 2004, Ren Zhengfei took me to the data center at Huawei's headquarters. I was shocked that he couldn't access the data center with his badge. He told me, "Not just the data center and the R&D center. I can't get into a lot of other locations because our processes don't give me the authorization." And I came to understand the awesome power of systems. At Huawei, the higher you are in the decision-making hierarchy, the less power you have in execution. The role of leaders is to build systems and to determine strategic directions. Generals then act on their leaders' orders and the actual battle is fought at the division, regiment, or battalion level. If the decision makers are too involved in execution, cliques tend to develop and corruption follows. Rules and procedures can eliminate a lot of these problems, but of course not all of them.

Ren Zhengfei argues,

> only by strengthening the negation of personal authority, and by strengthening our culture of self-negation, can we build up a healthy organization that doesn't depend on any one individual. When we have the right IT in place to handle management tasks, the sky won't fall no matter who leaves the company.

Ideological Criticism: A Cultural Password That Can't Be Copied

According to Mao Zedong, China has made three major contributions to the world: traditional Chinese medicine (TCM), the *Dream of the Red Chamber*,[3] and mahjong. TCM believes in the mutual restriction and generation of the natural elements, the complementary nature of yin and yang, which both attack and defend one another. Therefore TCM doctors are like philosophers and strategists because they have to attack certain things in the body while defending something else. The classic novel *Dream of the Red Chamber* is full of intrigue between men and men, women and women, and men and women, hundreds and thousands of words spoken with gentle feminine breath, not a single clink of dagger or knife—it's enough to suffocate its readers with beauty. And in the game of mahjong, each of the four players is like a country or kingdom fighting against the others, one against three. Each player has to play offense and defense at the same time to win the game.

These represent the national character of China. We are a wise nation—perhaps too wise at times. Therefore, Confucius demanded that his disciples reflect on conduct and consciousness three times a day and Mao Zedong called on everyone in the country to stir up revolution deep in their souls. It is clear that both Confucius and Mao understood the Chinese national character very well, including its merits and weaknesses, and this is why both great philosophers had proposed the same thing: ideological criticism.

Christian confession is similar in some ways to the self-reflection that Confucius demanded of his subjects, but it is very different from the spiritual revolution that Mao spoke about. Ideological criticism has reared its head in a number of political movements in Chinese history, the most revolutionary of which being the Cultural Revolution, which harmed and distorted the spirit of the entire country far more than words can ever capture.

---

[3] *Dream of the Red Chamber* is a widely celebrated Chinese classic, originally published in 1791.

In consequence, over the past three decades, ideological criticism has failed to make any difference in China because people instinctively fear and avoid it. At Huawei, however, it was picked up by Ren Zhengfei and has been used for over 20 years. Every department or division of Huawei convenes meetings of self-criticism and peer critique, and every meeting is serious. Each person is supposed to talk about their mistakes and problems without shifting blame or doing so in a perfunctory manner. They are also expected to find the mental cause of any wrongdoing. Other members help criticize in a constructive way, but the company forbids exaggeration, personal attack, moralizing, or any emotional outbreak.

As a result, individual and organizational problems do not plague the company for too long before they are exposed and corrected. Equally important, criticism and self-criticism have not ripped the company apart, but have created a dedicated team of more than 170,000 people. And that's not to say that the company has tamed its managers and employees into obedient, subservient yes-men; everyone has a unique, distinctive character. "The boss really doesn't like people without personality and passion. He doesn't really bother to talk to people who don't have passion, as he thinks a lack of passion is the most serious shortcoming someone can have. A waste of space," said a senior executive at Huawei.

Self-criticism is a tool, but it's important to ask: What should we criticize? How? And why? If it's poorly done, self-criticism won't unite people or increase organizational power, but will produce negative or even opposite outcomes. In China, meetings for criticism and self-criticism have been institutionalized in many SOEs and it's not my place to judge how effective they have been. Private companies, especially those owned by former military officers or government officials, have also tried to purge their companies through such meetings, but they have quite often ended in failure, with criticism leading to infighting and division.

Some companies, therefore, such as Hainan Airlines, have turned to religious faith to pull their people together.

As a Huawei senior executive remarked, self-criticism, especially ideological criticism, is really tough to handle. So how has Huawei managed it? The recipe is two-fold: The first ingredient is that the top management, who are not afraid of losing face, serve as role models, and the second is compromise.

This is very difficult for most companies to achieve.

It is tough for Chinese companies and even tougher for Western companies. In places with a strong and prevalent Christian tradition, many people have the habit of confessing to redeem their souls, but confession is rarely carried out between colleagues or organizations. Dissection of thought is an invention of China, where there is no predominant religion. In the West, ideological criticism touches on respect and privacy, so it's difficult to find any research on this topic in their management literature. The tendency is to analyze human nature from the perspectives of psychology and behavioral science, and most would rather depend on institutional measures or Western-style corporate culture to stimulate people's potential and suppress the uglier sides of human nature.

Huawei appears elusive to the West largely for its culture of self-criticism, reflected most notably in what the company calls "democratic meetings" or self-criticism sessions. Western companies will never really understand how it works, nor will they borrow this practice for their own use, because it is something that's deeply encoded in Chinese cultural heritage. In fact, not all Chinese companies are able to fully comprehend or utilize this philosophical tool of self-criticism; even if they could, they have a long way to go before fully Westernizing their management systems. Therefore, though many other private companies from China have tried to enter the global market over the past 10 years, it's not at all surprising that Huawei has been the only dark horse among them. That's not quite Chinese, but not quite Western, while at the same time a little bit of both, striking fear into the heart of many of its peers in the West.

Executive Declaration of Self-discipline: Two Sides of the Same Coin

The Maldives is a paradise on earth, where time seems to stand still. The beach, sunshine, seawater, blue sky, and coconut trees help tourists escape the hustle and bustle of the city and business world. It was here in

December 2005 that Huawei held its EMT's self-criticism session to discuss issues with integrity and self-discipline.

While it might seem odd for such a beautiful and relaxing island to be chosen for such a serious purpose, Ren Zhengfei and his colleagues were resolute and well-prepared.

All EMT members had agreed that, as key leaders of the company, they must be clean, upright, and set good examples for other members of the company. Therefore, at the meeting, they approved an EMT declaration of self-discipline, which demanded that all EMT members, as well as middle and senior managers, declare and clear their relationships with suppliers in two years' time. As part of this declaration, all managers would be required to take a public oath, putting themselves under the supervision of everyone in the company.

Two years later, in September 2007, Huawei organized an oath-taking meeting at its headquarters in Shenzhen. Face-to-face with over 200 middle-level and senior managers, all EMT members—including Ren Zhengfei, Sun Yafang, Guo Ping, Ji Ping, Fei Min, Hong Tianfeng, Xu Zhijun, Hu Houkun, and Xu Wenwei—raised their right hands and declared:

> We hereby commit ourselves to integrity, dedication, and diligence, by which we shall lead the company in its journey through the hidden perils and challenges that lie before it. We shall not allow leaders to set bad examples for their subordinates. We shall not allow the organization to crack from the inside out. We shall honor each and every one of our commitments under the supervision of the company, and all employees at Huawei.

After taking this collective oath, each EMT member proceeded to take an individual oath.

For readers to have a deeper understanding of Huawei's culture of self-criticism, what follows is the full text of Huawei's EMT declaration of self-discipline:

> Huawei bears a great mission and is committed to the common ideals of all its employees. For 18 years, we have contributed our

best and endured tremendous hardship to make the company what it is today. In order to ensure ongoing success, we must remain dedicated for decades to come.

We love Huawei in the same way that we love our own lives. To ensure Huawei's sustainable development and long-term stability, we shall heed the warnings of history so as not to fall prey to collapse from within. Hereby, we solemnly declare:

1. We shall set a good example of integrity and self-discipline. We shall derive all of our income from the dividends and compensation distributed by Huawei, and shall not attempt to obtain any income from any other source.

We shall not use the authority the company has given us to attempt to affect or interfere with the company's business activities or pursue any personal interests. Such business activities include, but are not limited to, procurement, sales, cooperation, and outsourcing. We shall not damage the company's interests in any way.

We shall not set up any company, or hold any equity or job in any other company. We shall not allow any company that is owned by our family members or relatives to enter into any related-party transactions with Huawei. We may help anyone we would like to help, but the money used for such purposes must come from our own pockets. Personal and company interests must be clearly separated.

2. We shall be honest and selfless. We shall not attempt to establish or lead any factions within the company. We shall not allow any improper work ethics to spread within our jurisdiction.

3. We shall restrain and discipline ourselves. We shall review our own conduct through self-inspection, self-correction, and self-criticism, and develop a self-cleansing mechanism for managers within the organization.

We fully understand that, as key members of the company, we must lead the company forward. Therefore, we shall remain united and focused on the company's development. We hereby commit ourselves to integrity, dedication, and diligence, by which we shall lead the company in its journey through the hidden perils and challenges that lie before it. We shall not allow leaders to set bad examples for their subordinates. We shall not allow the organization to crack from the inside out. We shall honor each and every one of our commitments under the supervision of the company, and all employees at Huawei.

Within a month, from May to June 2008, every department and subsidiary of Huawei made the same declaration.

Ren Zhengfei wanted to make the entire company completely transparent from top to bottom, to promote a democratic environment of mutual supervision. Some doubted Ren's intentions, arguing that he was managing the company by launching "mass movements."[4] This is an unfair judgment: Even if it was a so-called movement, if it proved to be a healthy and effective in the end, then why reject it so summarily?

Actually, this type of movement has occurred more than once in Huawei's history. Other movements include the mass resignation of the sales department in 1996 and another one involving 7,000 people in 2007, which not only produced good results back then but also laid an important philosophical foundation for Huawei's long-term organizational development, reform, and global expansion.

Objectively speaking, the classic doctrines, experiences, and solutions gleaned from Western organizational theories can no longer serve as readymade templates for Chinese companies as they progress further in the age of globalization. Based on a foundation of open-minded learning, Huawei has to reconstruct itself with its own unique ideas, and in doing so, look for methods and tools from China's own history of organizational governance. Obviously, a complete reconstruction would be preposterous, but blindly copying from the West would also be a fruitless effort. After all, Western companies are also still in their own process of exploring how to best manage knowledge workers.

---

[4] Translator's note: Here the word "movement" is being used pejoratively—the same way it was used earlier in this chapter, referring to various mass movements that Mao Zedong launched during the Cultural Revolution.

Having said that, in the end, any movement-like approaches need to contribute to the construction of systems. Systems help simplify organizations and also the relationships between units, teams, and individuals within an organization. A simple culture is the foundation of any organization's ability to remain strong and powerful. In this respect, Huawei has at its disposal a worthwhile series of established practices and systems that Eastern and Western companies have built over time, based on universal human logic.

Why did Huawei launch the "self-discipline movement" across the entire organization? The reason is simple. There are two extreme forces in human nature: the angel force and the devil force. The leaders in an organization are responsible for driving off the devil force so that people keep on plugging away at their work—to turn greed and sloth into the creativity that is necessary for the organization's development. To achieve this goal, the two sides of the coin—or words and actions—have to be totally consistent with each other, otherwise there will be chaos. An organization's ability to maintain strong morale from top to bottom revolves around the purity of its culture. If there were any discrepancies or contradictions between words and actions, especially between those of senior executives, Huawei would surely lose the cohesion and drive of its people. If employees were told to be customer-centered and dedicated, but they observed the company's leadership being self-centered, "plowing their own fields," forming factions, or leveraging their position in the company for personal gain, then Huawei would be doomed to failure.

Therefore, the declaration of self-discipline is a manifestation, a public banner that puts managers at all levels under the supervision of 170,000 people. Only if its mid- and high-level managers commit to working towards one goal and only deriving benefit from one source will this entire workforce of 170,000 apply its own strength to the company's pursuits.

Apart from supervision via public opinion (notably with the company's online forum, Xinsheng Community), Huawei has also developed a powerful institutional supervision system. It is one of cold deterrence, but also an armed force that can be deployed whenever appropriate.

This system is comprised of an inspection unit that monitors business activities that are still in progress, the auditing department that engages in post hoc supervision, and the Party Committee (known as the committee of ethics and compliance overseas) that ensures staff integrity and work ethics. The auditing department operates under the principle of "thorough investigation of every case." According to Wu Shuyuan, the department president, "While unspoken rules are prevalent in society, we must maintain an unshakable bottom line at Huawei: Everything has to operate in open daylight." In the company, no one, even Ren Zhengfei himself, has a driver of his or her own and no one flies first class on business trips (unless you pay for it yourself). These kinds of clear-cut regulations have largely prevented the spread of inequitable, unspoken rules within the company.

Nevertheless, Huawei still has its own fair share of problems. Corruption, embezzlement, and rent seeking still exist; some departments and individuals have been known to garnish their performance with fake numbers, profits, and contracts. Meng Wanzhou, Huawei's CFO, cautioned: "Huawei can't build itself on beer foam…." In 2014, a data-verification team was set up within the finance department to check and ensure the authenticity of all accounting information.

The Death of the King and Organizational Criticism

The Law of the Animal Kingdom: Tiger, Wolf, Ant, and Mouse

One afternoon in 1997, Ren Zhengfei met with an executive of the Hay Group, a global management-consulting firm based in the United States. Wu Chunbo, a coauthor of this book, had also attended the meeting. He recalled:

> During the meeting, they kept talking about animals. Ren said multinationals are like elephants, while Huawei is a small mouse.

Of course, Huawei is no match for an elephant, so it has to adopt the qualities of wolves: a keen sense of smell, a strong competitive nature, a pack mentality, and a spirit of sacrifice.

The animal world contains many sources of inspiration for human beings. How did the dinosaurs die out millions of years ago? Scientists have never come up with a definitive answer. Tigers are kings among animals; so are lions. Together, they once dominated the entire animal kingdom, striking fear into the hearts of all other animals. In recent times, however, both tigers and lions are at the verge of extinction. Their own survival is threatened. Pandas are even worse off: Their silly, helpless expressions speak to a very hard truth that the loss of wildness and adaptability is a disaster for all living things.

Wolves are the perfect combination of nobility and wildness, aggressiveness and pack mentality, vigilance and tenacity; unfortunately, they are under threat from human beings and their odds of survival are on the decline. I am convinced, however, that wolves still have the opportunity to conquer their environment.

"Live like cockroaches."—This comment from Terry Guo, president of Foxconn Technology Group, indicates an extreme sense of helplessness among entrepreneurs and, at the same time, formidable grit. Louis Gerstner, former CEO of IBM, became an US business hero for rescuing the "dying dinosaur." When he was trying to turn IBM around, he famously commented that "if the elephant can dance, then all of the ants will be forced from the floor."

Like IBM, most world-class multinational companies boast a long history, an immense size, a leading industry position, and an air of nobility. And from these, they have derived two common diseases: organizational fatigue and corruption.

Huawei shares many of these traits as well. With its more than 20 years of history and a team of more than 170,000 employees, Huawei has risen to the status of corporate nobility with its number-one ranking in the telecom-equipment industry. As early as 1998, Ren Zhengfei issued the following warning: "Recently, a handful of people at Huawei have shown signs of corruption. The company will not allow this trend to develop further. Otherwise, how can we accomplish anything as an organization?"

Apparently, Ren Zhengfei is afraid that Huawei might one day go extinct like the dinosaurs or sway along the brink of extinction like the tiger.

Fear leads to greatness. Huawei has lived with fear for 27 years. Besieged by Western companies and encroached upon domestic competitors, Huawei's adrenaline has been pumping nonstop as it continues to fight off intrusions on all fronts. Just think about it. When Cisco threatened that it was going to make Huawei "go bankrupt," how could Huawei not fight back at full force? When some companies openly scaled up their efforts to poach Huawei employees, how could Huawei do anything but strike back quickly in a widespread state of alarm?

Animals without natural enemies are often the first to go extinct, while those challenged by their natural enemies tend to proliferate and prosper. This natural law applies equally well to the human race and the business world. While it grows, any business organization will meet with many opponents. They may pose various threats to the organization, but at the same time, their constant challenge drives the organization forward. When competitors or enemies are standing right in front of you, you can't afford to lose a bit of your aggressiveness or enthusiasm and the organization can't afford to lose even the tiniest bit of cohesion, nor the reason and dedication of its leaders. Therefore, a wise entrepreneur would never curse his opponents, but feel indebted to them for steeling his nerves and strengthening his organizational muscle, thereby enhancing his battle readiness.

Quite often, a great company isn't the product of ambitious strategy, but of a major crisis. This is at least true of Huawei. If Cisco had not sued Huawei, Huawei might not have transformed so quickly from a clumsy boxer into an adept tai chi master that knows how to balance between competition and cooperation, offense and defense. Similarly, a series of internal struggles since 2001 have triggered a change in the way that Huawei manages the company and develops its culture and systems, which has indirectly prompted its growth and progress.

One day in 2002, the leaders from Huawei's six major research centers in China met in Beijing to report on their work to Ren Zhengfei. When they broached the topic of a company that had only been established less than a year prior and how it would one day pose a threat to Huawei, Ren was livid. He grabbed the printouts on the desk in front of him and ripped them to shreds. What was going on in his head? Was he so furious because he didn't accept their take on the situation or was he angry about the fact that a new challenger had suddenly risen up in their midst? I can't say for sure.

However, what can be said for certain is that, after this, Huawei's leadership began taking precautions against any "questionable-looking moles" that popped up on the surface of the industry, setting up institutional operating tables should the need to lance them arise.

Huawei was facing the Cisco onslaught during this period. From 1999 to 2006, the company was stuck in a rut. Ren Zhengfei had grown so tired that he fell ill. He had two cancer operations and suffered from depression.

Psychologists argue that many forms of depression not only bring torment, but also sharpen one's mind, making people more aware of the problems of the world and more able to confront them. Many forms of mania can toughen people up, helping them to learn from past failure. In some instances, psychological problems are early signs of exceptional accomplishment to come.

This argument certainly applies to Ren Zhengfei. He once said, "We should not be resistant to criticism from any opponent. We must listen carefully, even if they have gone too far. To beat us, they need to find our weaknesses. As for us, they can unearth the problems that we're blind to." A colleague agreed and said, "A crow might be noisy, but no one has ever died from listening to its caw; an owl may disturb the night, but its cry warns of death nearby. In this sense, both the crow and owl are auspicious birds."

Ren Zhengfei nodded in agreement.

He divides self-criticism into two categories: ideological criticism and organizational criticism. Ideological criticism probes into questions on the correctness of individual and organizational values, whether they've been warped, and whether or not they foster a positive and reinforcing environment. Have individual or team work ethics grown corrupt? Have they mutated? These are the questions that ideological criticism addresses, and that are most frequently discussed at the executive level; these discussions then cascade down throughout the organization via self-criticism sessions. On the other hand, the company has remained more cautious and serious when dealing with organizational criticism, because when it reaches a certain point, this type of criticism involves structural reforms and taking action. "Transformation is a systematic undertaking, and can't be taken lightly." On this subject, Ren Zhengfei said the following:

> We have grown from a small company with scarcely any management system. After a series of transformations, including integrated product development (IPD), integrated supply chain (ISC), and four financial unifications, our management system has evolved from simple and rough to IT-based and fine-tuned. And our company is globalizing. Without self-criticism, no one would go near the prior policies or statements made by a leader, or revise a process that would lead to the removal of a certain position. But if we don't ever change anything, how could we build a complete system of business processes? Without these major improvements in our management, we would not have been able to provide lower-cost and higher-value services to our customers. In this competitive market where prices are bottoming out, we would not be able to survive. Our management system is subject to non-stop self-criticism, as without it we can't survive in this fast-changing world.

There was once a popular book called *Small Is Beautiful*. Huawei's senior executives argue, however, that a small piece of grass might be crushed at any time by a big elephant. This is an age of risks, and bigger ships can better survive a tempest. And of course, as Wang Xifeng said in *Dream of the Red Chamber*, "Being big has its own problems."

Huawei has grown itself into a large company through constant ideological criticism and cautious but deliberate organizational criticism. But in 2008, Ren Zhengfei said, "We shall now shift our focus from ideological criticism to organizational criticism." However, his target this time is one of the most serious diseases that large corporations catch: organizational fatigue.

Can Good Genes Guarantee Ongoing Success?

According to the economist Joseph Schumpeter, all successful busines-smen stand on ground that is "crumbling beneath their feet." The reason is simple: Fear leads to greatness. When the leaders and the whole organ-ization are under intense pressure, the organization's synapses will fire out alarm signals. These awaken everyone's collective instinct for defense and then turn that instinct into an aggressive attack force, significantly enhancing the organization's ability to fight off infection and disease. Homogeneity and exclusiveness are common characteristics of the culture in these types of organizations.

A factor that can't be ignored is that, after the successful conclusion of an event or when what appears to be a crisis abates, the organization will present an energetic and healthy front in all aspects and this is when people start to loosen up and relax. With that, the organization's immune system will begin to weaken and then its body will begin to experience the progressive, irregular occurrence of disorder.

The infamous Telekom Malaysia incident is a classic example of this loosen-up-and-relax syndrome at work.

In 2009, with record annual sales revenue of US$30 billion, Huawei became the world's second-largest supplier in the telecom industry. On August 5 the following year, Sun Yafang, Huawei's chairwoman, received a letter of complaint from the CEO of Telekom Malaysia. It talked about how, in the several months prior, Huawei's performance was not living up to the professional standards of a large international company. The letter further went on to describe multiple issues that had aroused great concern within Telekom Malaysia's management team. Specific issues included:

1. Contract compliance (product specifications) and delivery: Some of the delivered equipment did not match contract specifications and testing requirements

2. Lack of professional project-management practices (approaches)

3. Lack of expert resources specified in the contract

This letter of complaint "conveyed disappointment and anger between lines of polite expressions."

The letter was simultaneously sent to the president of sales and services, the president of Asia-Pacific operations, the president of the South Pacific region, the head of Huawei's rep office in Malaysia, and related staff in the department handling the Telekom Malaysia account inside Huawei. However, five days after sending out the letter, the customer still hadn't received a response from Huawei: The focus of people's attention at each level of management was not on how to address the issue, but on how to reply to that letter.

Five days later, when Sun Yafang returned from a business trip overseas, she was informed of the complaint letter. She immediately began going over the entire process and reflecting upon what had occurred in the Telekom Malaysia project.

What she uncovered indicated that Huawei had developed the following symptoms of the illness that affects many large corporations:

1. A focus on contracts instead of delivery: "Five years ago, Huawei's share of Telekom Malaysia's business was less than 10 percent of their biggest supplier's, but by the end of 2009, Huawei's share was 10 times greater than theirs," which made Huawei Telekom Malaysia's most important one-stop solution provider. However, "there was no process for solution delivery," which resulted in one glitch after another in the course of delivery.

2. Silos between departments, with no collaboration, blaming each other when problems occurred.

3. Overpromising to win contracts, then under-delivering or holding back the truth when products couldn't meet requirements, so that "the customer had no trust in us at all."

4. Red tape: emails sent back and forth to evade responsibility instead of immediate on-site inspection after problems occurred.

5. Passivity and bureaucracy at the senior-management level: During a one-year period after winning the bid to implement the Telekom Malaysia project, at least 10 VPs at Huawei (vice presidents,

generally referring to senior managers) were directly or indirectly involved, so there wasn't a lack of attention to this project, but a lack of communication among high-level executives, which led to insufficient resources, a lack of transparency around issues, two delivery failures, shipping the wrong equipment, and so on.

6. Concealing bad news: both from the customer and from their superiors in order to get away with problems or delay the inevitable.

It was also clear that the Telekom Malaysia incident was not an isolated case. It reflected a set of symptoms that had become common in Huawei after it had entered the ranks of the leading companies. As a matter of fact, according to a former senior executive, around 2006 or 2007, Huawei was already showing signs of laziness and complacency after its sales revenue hit US$15 billion. Tao Jingwen, president of Huawei's Western Europe region, once quoted the CTO of Vodafone as saying: "I don't know where Huawei got so many arrogant and self-satisfied young people all of a sudden."

Huawei made countless mistakes throughout its history, and was blamed and criticized by its customers on numerous occasions. For some time, the roadmap of its R&D organizations was too technology-oriented, not necessarily reflecting the actual needs of Huawei customers, which caused huge losses and strategic confusion in the company. That's why a quality reflection conference was held in 2000, at which some scrapped materials from poorly managed R&D projects were distributed to R&D staff, giving them the opportunity to reflect upon it and redeem themselves—to crawl out of the proverbial mud pits.

In spite of all this, in the eyes of its customers, Huawei remained humble and responsive, fully committed to customer service. Therefore, even if its products were not fully satisfactory and its technical support was rather weak, it still won their trust and recognition.

However, in the Telekom Malaysia incident, the customer's response was very different from what Huawei had ever experienced before. "I believe in your ability to solve the problems, but what I want to see is an attitude of willingness to resolve them… to see if you are listening attentively to what we have to say," commented the CEO of Telekom Malaysia. A vice president who was about to retire from Telekom Malaysia complained in dismay that "your [Huawei's] failure is likely to ruin my career."

The Telekom Malaysia case prompted Huawei to launch a sweeping campaign of organizational criticism. The entire management team was extensively briefed at an EMT meeting, then workshops were held across product lines to explore how to collaborate and harmonize internal operations. In field offices at both the country and regional level, they organized thorough self-reflection sessions to dive deep into the issues at hand. They tried to identify problems with existing systems and processes, but more importantly, the self-examination was extended to the ideological level: What is our relationship to customers? What is our attitude towards problems? Are we familiar with our customers' expectations of us? How do we achieve customer centricity? What does "dedication" mean?

With serious attention from Huawei's senior management, the Telekom Malaysia project was successfully delivered in the end. However, the company did not stop its critical reflection on this matter.

Xu Zhijun:	"As we work with more customers, we've begun to change our attitude towards them. When we were small, we treated our customers with respect and awe, but that feeling's utterly gone. These days, some people even joke about the problems we're dealing with."
Xu Wenwei:	"Another thing I've noticed is that we don't listen to customers' needs. Rather, we are self-centered, always trying to persuade customers to accept our products the way they are."
Li Jie:	"I recently visited six different customers, and they told me that each meeting with Huawei people is like going to class, where the teacher tells them what to do."
Li Shanlin:	"… it's critical that the entire organization become customer-centered, not leader-centered or self-centered. Systems and processes prevent people from making mistakes, while culture and values make people *unwilling*

to make mistakes. We have to think about 'the soft side of things,' especially when it comes to the inheritance of culture and values."

The Telekom Malaysia incident was later compiled into a case study of more than 10,000 words, "Are We Still Customer-Centric?—The Story behind Complaints from the CEO of Telekom Malaysia," and was published in the journal *Huawei People*. The authors were senior executives, including Xu Zhijun, Xu Wenwei, Ding Yun, and Yao Fuhai.[5]

The case study closed with a penetrating question: "If we are customer-centric, we can foster genius; if we are leader-centric, we become underlings; and if we are self-centered, we are nothing but fools. People of Huawei, which do you choose?" It also included a quote from Ren that Huawei's boom had left it at the brink of going bust.

Fate and Mission: When Will Huawei Die?

Throughout China's 5,000-year history, only four dynasties were able to last longer than 200 years. The average life span of the top 500 companies in the United States is 30–40 years, and some have drawn the inference that the life expectancy of a company typically tops off at about half that of a human. Only one-fourth of the companies on the Dow Jones Index have survived more than 50 years.

A Japanese management expert proposed a measurement of time: dog years. One dog year is usually considered the equivalent of about 10 human years. (Writing this, I suddenly felt a bitter pang of anguish as I looked down at the three dogs curled around my feet: They are so cute, and you mean absolutely everything to them. One day they will leave me, and when that day comes, I don't know if my heart can take it.) And if that's the case, then most Chinese companies have an even shorter life span: less than one dog year!

---

[5] At the time this book was written, the respective roles of these executives were as follows: Xu Zhijun was deputy chairman of the board and rotating CEO. Xu Wenwei was president of the strategy marketing department. Ding Yun was president of products and solutions. Yao Fuhai was president of the procurement qualification management department.

For over 30 years after reform and opening up, China has produced many excellent companies that were once the focus of MBA case studies around the world, but now 85 percent of them have fallen to the wayside. By 2010, the average life of Chinese SMEs was 3.7 years, while the average lifespan of European and Japanese companies was 12.5 years; US companies survive 8.2 years on average and about one-fourth of the 500 best German SMEs have survived more than 100 years.

This is the fate of every organization: they will all die, sooner or later. And the mission of their leaders is to lead every member of the organization in the fight against this eventual demise; they must struggle to live longer and healthier.

As a materialist, Ren Zhengfei is not afraid of talking about death. He said, "Life always ends. What we are doing is trying to make Huawei live as long as possible. We don't want it to die off so soon, and we don't want it to die in a miserable way, either." To him, the most basic strategy at Huawei is to live on.

On January 5, 2012, an Argentinean media website posted an exciting scientific discovery from in article entitled "We Are Getting Closer to the End of Death." The news read:

> Modern biology has proven that adult stem cells do not age like other cells in the body. That means that before most cells start to age, the stem cells will function regularly. Now we know that stem cells do not age, they can be frozen and kept forever, and then thawed when we're ready to use them.

The author went on to predict: "Some scientists are studying the aging process as if it were a disease, and perhaps in 20 or 30 years we will witness the beginning of the end of death. The dream of living forever is getting closer and closer to reality."

This news is thrilling, no doubt, but living forever is still far away from becoming a reality—certainly farther than what the author had predicted. Nevertheless, this idea triggered some interesting associations: What and where are Huawei's stem cells? What is Huawei's worst nightmare?

Without a doubt, Huawei's stem cells are its core values: customer centricity, dedication, and perseverance. Customer centricity and dedication have combined into an embryo that has grown into the company that we call Huawei. And perseverance is its most prototypical gene. Huawei would not have survived without this gene, which has remained unchanged for over 20 years.

Many people, including those both inside and outside of Huawei, have associated the fate of the company with the fate of its CEO, Ren Zhengfei. This is a mistake. One day Ren will undoubtedly have his curtain call, as this is an immutable natural law, but the company itself might not meet a similar end. The overall numbers do not look bright, with the "dog years" argument and an estimated average life expectancy of 3.7–12.5 years. But there is such thing as a company with a long life. In fact, people's understanding of the organizations they create is still superficial and limited, and has yet to break the surface of organizational life cycles.

As for Ren Zhengfei, the limits of his professional life are relatively predictable. His requirement for members of his executive management team is to work for the company in a healthy manner for 20 years. And for himself? Well, he is already 70 years old. However, the organization he created together with over 170,000 people might last a century or even longer—while on the flip side, it's also possible that Huawei will only survive for another 10 years, even less. Outside of external factors beyond its control, such as war and social unrest, Huawei's own problems, such as aging, decay, and middle-age obesity, are far bigger challenges.

Therefore, the key to survival is whether the company's stem cells, so to speak, can continue to function with vigor and replicate themselves in every system of the organization. In other words, Huawei will die if it loses its stem cells, or its core values, no matter how long Ren Zhengfei stays around to lead the company. On the other hand, if it keeps its core values healthy and robust, never letting go of its spirit of perseverance, Huawei will thrive and keep growing long after Ren Zhengfei leaves the helm.

Can you imagine an elephant dancing? An elephant is big, but that doesn't mean it's clumsy. Life exerts itself through movement, and Huawei is still recruiting new employees and expanding its organization day by day. Today, it has 170,000 people, and in 5–10 years' time, that number might climb to 200,000 or 300,000. The decision makers of the company will hold fast to its core values and keep engaging in a "dissipative exercise" in order to stir up vitality within the organization. Recently, the company has adopted some terms from the military in its internal communications: "Let those who can hear the gunfire call for artillery support"[6] and "battles commanded by field squad leaders,"[7] among others. They represent the insight and reflection of Huawei's leadership, garnered since 2008, into the diseases of large corporations.

Appendix: The End of Competition Signals the Loss of Momentum

Management Insights from Redwoods and Tropical Rainforests

1. There are two kinds of trees known to be the tallest in the world. One is the redwood that grows in the suburbs of California and the other is the mountain ash (*Eucalyptus regnans*) in Southeast Australia. Both can grow to approximately 150m. There's still no conclusion as to which is superior, but the two have some things in common:

    i. They always appear in large communities. Redwood forests usually stretch hundreds of kilometers and mountain ash are also gregarious in nature. It is difficult for individual trees to

---

[6] This means that frontline employees will be provided with full support from "the big guns," or technical experts, at headquarters, and will be equipped with the means to call for this support as needed.

[7] Similarly, this phrase refers to frontline employees receiving more decision-making authority, so they can be more responsive to market conditions without relying on headquarters for permission.

survive natural disasters, so the area covered by the group in a sense determines the height of its individuals.

ii. They have a near-perfect structure. In addition to a straight trunk, a large and thick base, and roots that stretch out both wide and deep, even the leaves of mountain ash have evolved to face the sky sideways, parallel to the direction of sunshine. In this way they have adapted to the dry climate and strong sunlight of their environment by avoiding direct solar exposure and excessive evaporation.

iii. Their height has a definite limit. Despite the aforementioned conditions, no samples taller than 200m have been found in vegetable history. Logically, this doesn't make sense. With their perfect structure and a high tolerance for variance in sunlight and moisture conditions, redwoods can live up to 2,000 years and yet they can't break this height limit.

2. Biologists are particularly interested in this inability to grow beyond a certain point and have expressed a variety of opinions on the matter, among which two views have reached rather wide-spread consensus.

i. Their perfect structure means a loss of efficiency. One reason that the trees can't grow taller is because they are already tall enough. Australian mountain ash have highly developed root systems, fairly exquisite capillaries, and perfectly arranged tree trunks. The result of such a combination is that, on average, it takes 24 days for a drop of water to be transported from the bottom of the roots to the leaves at the top. The optimization of each part leads to the lack of a better solution for the whole.

ii. The end of competition means loss of momentum. Another reason why the trees can't grow taller is simply because they are already the tallest among all plant species. The plant community has reached a subtle balance over many years of evolution. Other plants are no longer able to compete, and within the community, all other plants have accepted this fact. Entropy always increases in a closed system, so it's not that some trees have stopped growing, but that a group of trees has stopped competing.

3. Philosophers and physicists have actually travelled a long way along this path of thought. The second law of thermodynamics proposes a thermodynamic arrow, indicating that the universe began expanding immediately after the Big Bang and that everything will end in a disorderly state of heat death, so time is irreversible. Or in more plain terms, an increase in disorder or entropy is the norm as time moves forward.

4. Therefore, for a single organism, the ultimate question is how to introduce negative entropy and whether it's an act of evolution or of adaptation. This has been a controversial issue among biologists for nearly 200 years. Many people confuse the two. "Adaptation" is the practice of a single generation whereas "evolution" stretches across many generations. Adaptation is the foundation of evolution, and evolution is a process of adaptation. No matter how adapted a single organism is, it will still eventually meet its end. In a universe full of disorder, the orderliness of an individual must be built on generations of collective evolution. However, the direction of evolution has to be adapted to the individuals and the environment of each generation (a larger system with more individuals).

5. Following this logic, biologists have found a counterpoint to the redwoods—the tropical rainforest. The Amazon is a very sophisticated prototype of an open system and it has recently become the focus of research by scholars on management, Internet-related industries, and futurologists. Indeed, many aspects of the rainforest are worth analysis and observation. Let's look at two examples.

    i. *Orderly disorder*. The Amazon is the epitome of species diversity in the world, where all manner of organisms can find their own place. Behind its seemingly chaotic and uncontrolled development is the precise logic of survival: the nonstop, automatic adjustment of individual adaptability based on the pressure to survive, which is passed along by the food chain itself (i.e., the order and hierarchy of evolution). Alien and underdeveloped species either adapt themselves to this system by adjusting their positions in the chain or they are eliminated from the system entirely. There are no supreme beings or CEOs controlling any of these activities. People like Darwin have discovered with great awe that this immense self-operating system has its inherent laws.

ii. *Long-term group evolution.* An individual tree can never avoid the disease and destruction brought about by the thermodynamic arrow, but the rainforest can. A group of redwoods can't grow taller, but a tropical rainforest can go on forever. Individuals pass down their "adaptability" to the next generation through their genetic code. Each improvement is very small, but it can be stabilized through countless generations. With neither boundaries nor passports, the Amazon community remains open, compromising, and gray. Single-gene mutation, alien-species invasion, and environmental change can all be addressed through the selection and adjustment of the group's genes. Those that introduce negative entropy are definitely open groups.

(Quoted from the reading notes of a Huawei manager)

On Change (I): Seven Prohibitions and Eight Symptoms

Chapter 7

Facing the Cold Sharp Blade: Change or Die

Metamorphic Redemption: Shaking Off Dependence on Heroes

Over 2,000 years ago, Confucius, a frustrated philosopher, stood on the banks of the Yellow River and sighed, "Time is passing like this river, flowing away endlessly day and night."

Two thousand years later, on April 28, 2005, a man named Ren Zhengfei was comparing himself to a fleeting sage who, in a revolutionary act of self-exile, would shatter his own glory and create a brand new future for the company.

On this day, Ren Zhengfei was invited by the CPC committee of Guangdong Province to give a speech on Huawei's core values. He said:

> Managing a company is like managing the Yangtze River. We build the dam, and the water flows through it, day and night. When it's nighttime and I'm asleep, the water still flows. It flows into the sea and evaporates into a gas, later to fall as snow on the Himalayas. From there it melts back into water and returns to the Yangtze River. The cycle is completed when the water flows into the sea to evaporate all over again. After many cycles like this, the water

has long forgotten the sage who stood on the bank of the river, mourning the passage of time. All it knows is the flow, so it keeps flowing. And who's that sage? Well, the founder, of course. A company is at its liveliest when its founder is no longer useful. So when the founder stands in a prominent position of prestige and is worshipped by everyone in the company, that's when the company is at its most hopeless, when it's in the most danger.

Some media outlets, both at home and abroad, have doubts as to Huawei's fate once Ren Zhengfei leaves the company. If you take a step back and look at the situation, when the founder of any preeminent company leads the company to its peak, it's done at precisely the same moment when that founder has reached the twilight years of his or her own life, at which point the media and public as well as the company's competitors will sigh over the fate of the company and begin making all manner of alarmist speculations.

Ren Zhengfei, accustomed to self-criticism, is keenly aware of this phenomenon. A scholar who has served Huawei as a management advisor for many years said:

> Ren Zhengfei is different from most other entrepreneurs. Company founders are generally attached to their company, both physically and emotionally. They believe their company is their baby, born after a long pregnancy and raised up through enormous effort. Now a growing child, the company has become a part of their life, and naturally they won't let it go. Ren Zhengfei is an exception. He believes that a company has its own independent life after it's born. The founder at some point will leave or die, but the company and its management system can regenerate itself and last. In this sense, no individual should be allowed to interfere with the course of its development. At a certain point, a company must cut the umbilical cord—that emotional link with its founder.

The transformation of Huawei's management system was just such a turning point, waving its cold sharp blade to sever the link.

The year 1998 was a critical milestone for Huawei. Since then, the company has systematically introduced the advanced management experience and practices of its management advisors. Through partnerships with leading consulting firms, including IBM, Hay Group, PricewaterhouseCoopers, and Accenture, Huawei built up its own customer-driven processes and management systems for IPD, ISC, human resources management, financial management, marketing management, and quality assurance.

This was a massive tectonic shift for Huawei. As Xu Zhijun and Guo Ping put it, this was an organizational revolution. Powerful people lost all or part of their power and the exercise of power became subject to restrictions. Furthermore, the leader of the revolution would have been overthrown if the change had not been carried out well. In history, most champions of reform came to a miserable end because they harmed the interests of too many people. When Huawei became determined to replace personal governance with procedural and institutional governance, it had to pay the price as well.

For 17 years, Huawei has consistently engaged in management transformation across six major dimensions of its business. Throughout this time, over 100 mid-level and senior managers have left the company or have been demoted or replaced because they resisted or could not adapt to the change. Many of them were extremely capable people or had made significant contributions to the company. But the new process transformation was the company's chosen path and Huawei no longer needed individual heroes in its management team.

In the beginning stages of its IPD transformation, Ren Zhengfei was known to say the following:

> If there ever comes a day that Huawei burns to the ground, every one of you would pack up hearth and home[1] and get the heck outta Dodge. But as long as our systems and processes are still intact, we could still build up another Huawei. The reason why we made such a huge decision to transform the company, and why we have spent so much money on hiring Western consultants to teach us the ropes, is because we want to cast off our dependence on individual heroes within the company. Among

---

[1] Translator's note: Here, Ren actually (and humorously) quotes the lyrics from a famous folk song, "The Girl from Dabancheng:" "Pack up your dowry, and take along your little sister."

regular employees, we still advocate the existence of heroes. But not mid-level and senior-level managers; they can't be heroes.

In a sense, heroes are often the stumbling stones along the path to reform. When a company starts from nothing, heroes are necessary and they help the company with their glorious deeds. Every company has gone through such a period of heroism. Heroes create history, but in most cases, with their ambitions and desires, they can also make history a dirty and bloody mess. Therefore, when a start-up company reaches a more orderly and mature state, heroes must be removed to pave the way for reform.

No doubt the biggest enemy of transformation is people, especially leaders. Any act of reform would be easier and more effective if people yielded to it and not vice versa; otherwise, it takes twice the effort for half the results. To some extent, a reform is a type of revolution because both of them abolish and redistribute power. There are some differences, though. A revolution is often violent and ends in a regime change, whereas reform is much more reasonable and more constructive. Reform aims at stimulating vitality within an organization, overcoming laziness and corruption, and optimizing organizational structure.

Unfortunately, in many companies, when they reach a certain stage, owing to outdated modes of thought, the leaders prefer securing their own interests and power through violent movements and interpersonal conflict within the organization. This type of reform never serves to push the organization forward. Instead, an organization is far more likely to tremble and even fall apart under such percussive duress.

Huawei's management reform covers every person in the organization, including Ren Zhengfei himself. Ren once admitted:

> In Huawei, the top leaders have less power and command fewer people than lower level managers. For example, a rotating CEO has only an assistant and perhaps a secretary, and does not possess any direct command authority. The chairwoman and I are symbolic leaders. We have no decision-making power. We can only veto decisions or impeach other leaders, but we have rarely exercised these rights. They can't be used often, no more than two or three times a year, or they would lose their effect.

It's like an atom bomb: they provide deterrence as long as they are not set off. On the other hand, our leaders in sales and R&D functions command 40,000 or 50,000 people. They have a huge amount of power. Of course, we have instituted a mature supervisory system to oversee them.

Does time flow like a river? Yes, it most certainly does! As Ren Zhengfei said, "Huawei's transformation is a process of dissipating individual authority. Huawei will become truly mature when it does not depend on the influence of a handful of people."

Mindset: The Worst Enemy of Change

China has a long history of raising birds, while the West takes fancy in training and riding horses. Birds have little freedom in the cage and are dependent on their master for food and water; in return, the only way they can express their appreciation is by singing for their master—in the end, they aren't much more than a plaything. Training horses is different. Horses are accustomed to running wild on vast land. People often compare unrestrained imagination to wild horses running free. To train a horse, one must allow the horse to retain some of its wild nature while attempting to regulate its behavior. In other words, a well-tamed horse runs with almost warlike passion but within a certain set of boundaries. Compared to birds, the world of horses is so rich and fulfilling! And from this it's clear that the cultures of those who raise birds and of those who train horses are two worlds apart. What sort of enlightenment can corporate managers gather from these differences?

A birdcage represents a certain set of rules and prohibitions. And in a vast land where horses can run wild, although there is a fence around the outskirts of the property, there is also freedom and imagination within. In the minds of most company managers and employees, rules are equal to restrictions, and one of the mistakes that company managers make is to pit rules and goals against each other. They falsely believe that rules will make goals impossible.

Ren Zhengfei said, "Huawei's most basic goal is survival, and of course we will try to catch up with our Western counterparts in the long run. To achieve this goal, we must put on our American shoes." Zhang Zhidong, a Chinese politician and reform advocate during the late Qing Dynasty, argued that "we should depend on the Chinese tradition while drawing on achievements from the West." From a practical perspective of national reform, it was clearly a failed act of cultural and ideological adoption. But Ren Zhengfei comes from a different time and Huawei is at the forefront of globalization, so he naturally has fewer restrictions in the way he thinks and a broader awareness of history and the world.

Huawei's "American shoes" refer specifically to a set of corporate governance systems that are well established in the West. In Chinese terminology, they are the Tao, or the way, and at a much deeper level, they also include specific techniques. In order to enter the West, Huawei had to completely Westernize its management systems and, in terms of management philosophy, it had to find a way to mix Western concepts with Chinese ones. And if the company could manage to keep this up, it would definitely set Huawei up for long-term prosperity.

Therefore, in 1998, the institutional and procedural reform that Huawei set in motion had deep implications. It was intended to rectify the "pirate culture" that had developed in the company's formative years of hunting and gathering, and would certainly meet with resistance from people who had vested interests in that culture. However, the more difficult part of this reform was the collision and integration of the different cultures, ways of thinking, and disparate personalities between the East and the West.

In a sense, the worst enemy of change is people's mindsets, followed by vested interests. John Maynard Keynes said: "I am sure that the power of vested interests is vastly exaggerated compared with the gradual encroachment of ideas. [...] But, soon or late, it is ideas, not vested interests, which are dangerous for good or evil." And this is true, at least in the case of Huawei. Was there anyone who opposed reform within Huawei? Certainly. And many were removed from their posts as a result. But at that time Huawei was only 10 years old with a generally young workforce; they didn't pursue vested interests in a systematic way, so organized opposition against the company's transformation efforts never surfaced.

For Huawei, therefore, the hardest nut to crack was reforming people's mindsets. If their mindsets were rigid and arrogant, their resistance to reform would be intangible, dense, and ubiquitous. For example, a conceited R&D director might think that, "Well, my way has led to success—why change it?" Or pessimists might look at all the company's efforts to reform and think that the probability of failure was greater than the probability of success; they might think that spending a ton of money on reform would drag the company down "and then everything I've achieved over these past 10 years will be gone." As Xu Zhijun puts it, "There have been a lot of pessimists at Huawei over the past 20 years, whereas our boss believes in a crisis doctrine. But he's not a pessimist; at his core, he's an optimist."

In the initial and middle stages of reform, Ren Zhengfei was always trying to "brainwash" pessimists with the idealistic expectations that accompany transformation. At the same time, he issued his "cut the feet to fit the shoes" imperative, calling on people to adapt or be cut off. Only in this way could they put an end to all the senseless debate and achieve consensus on reform across the entire company.

When he mobilized the core team members of the IPD transformation project, Ren Zhengfei was stern, even harsh, but spoke with the deliberate and convincing cadence of a teacher:

> First, we need to dismiss outright the people with half-baked knowledge who hop on the reform bandwagon for attention. We must also eliminate the lazy and unambitious. We want every one of you to wear a pair of American shoes, and we will have our American advisors tell us what American shoes are like. You may wonder whether those American shoes can be adapted a bit after they come to China. Well, we have no right to change anything; that is at the discretion of our advisors.
>
> From now on, you'd better not raise any new suggestions to show off your own talent; unorthodox ideas will not be rewarded. Instead, we must try to understand the substance behind IBM's management system; those with the most profound understanding will be rewarded. Some may ask whether they can make any changes once they understand the system well enough. I would

say yes, but not now. Maybe 10 or 20 years from now, but for the time being, you don't have this authority.

I believe that, right now, we need an extremely rigorous method of study. Innovation must be based on proper understanding. There is no point in expressing a new idea before you have fully understood it. That's seeking the limelight, and I think we should remove any such people from the core team.

We should remove those who can't understand the IPD reform, too. Membership in this core team is not lifelong. I wonder if we can't renew membership every month. We'll publish the new assignments on a monthly basis. I am happy to sign my name 12 times a year—and I'll even do it free of charge. This core team should have a revolving door with people coming and going. We must not give all the opportunities to those who are looking for a free ride on the bandwagon, or the intellectually lazy.

A lot of junior managers have complained that Huawei is not fair. I asked them why. They said that you, the senior managers, have grown up with the company's resources, and if they also had the opportunity to attend those expensive training sessions that the company paid for, they'd be just as good—even give your vice presidents a run for their money. I think they're right. Many of you have been to those training sessions with our outside consultants, but you've shown no improvement. If you don't improve, then you can get out too! Each time we appoint new members to this team, we have to take out one or two existing members, even if everyone is doing just fine. For people who have a head on their shoulders and are eager to learn, we'll let them in, let them compete. If the people in this core team aren't performing up to snuff, then we'll replace you with them. If we're going to learn from them [the West], then we need to fully commit ourselves to it. We need to understand the full picture of what they're telling us, and not just grab on to one specific point.

Of course, there are many good management practices out there, but we can't embrace them all. That would turn us into a bunch of idiots: Different management approaches lean in different directions, and trying to accommodate them all would simply get us nowhere. Therefore, we will only learn from one advisor, and learn a single model. In 10 or 20 years, when we are a US$200 billion company, we can perhaps embrace something new. At the moment, Huawei hasn't even learned a single model well enough, so who are we to come up with something new? Our transformation efforts over the past couple of years have failed because we kept on trying to come up with something new and fancy on our own, and nothing works. This time we must stay level-headed, focused, and pragmatic—we will put on this one single pair of American shoes. We must humbly learn from the best if we are ever going to beat them.

The management reform led by IBM has lasted 17 years at Huawei. At any given point in time, there has been anywhere between a dozen and a hundred IBM experts on Huawei's campus, each charging US$300–600 per hour. This is perhaps the highest price that any Chinese company has ever paid for business consulting services.

And what are the results?

In 1997, IBM did a comprehensive review of Huawei's management system and found loads of problems. The company wasn't able to accurately assess prospective customer needs. It was repeating useless work and wasting resources, which drove up costs. While processes existed at the department or unit level, there were no structured cross-functional processes, nor well-defined and automated process integration. Strong organizational silos caused internal friction. There was a lack of professional skills, tasks were non-standardized, and the organization depended heavily on individual heroes whose success was not easily duplicated. Project planning was inefficient, project implementation was chaotic, there was no change management, and version control was a catastrophe.

Imagine what the company would have become if it had not carried out a fundamental institutional reform. In Ren Zhengfei's mind, he imagined ruinous collapse—sudden ruinous collapse, like an avalanche. And so it was that he himself was under crushing amounts of pressure.

Over 10 years have passed, and Huawei has not collapsed. Instead, it has become a formidable competitor in the telecom world. IPD reform has been vital to the company's success. As a modern management system, IPD more systematically standardizes the company's stops and goes, what it forbids and what it lets go: The land is vast, but it has boundaries; the river runs fast, but it runs through a solid dam. In other words, in the new system, the customer demand sets the direction and the process framework sets the boundaries; business operations are made efficient to ensure high-quality delivery, end to end.

Even with small internal earthquakes from time to time, the IPD reform has proceeded calmly and quietly at Huawei. It has achieved initial success, and the reform will continue on.

What Huawei Learned from the Whiz Kids

On June 16, 2014, the first "Whiz Kids" award ceremony was held in Huawei's headquarters in Shenzhen. Prizes were awarded to more than 100 winners from middle to top management. Most had led and participated in transformation efforts across different domains of the company. Among them, there were also the IBM consultants, Arleta Chen, Mee Wong, and Lew Kimmel, the three key advisors for Huawei's IPD transformation.

The Whiz Kids award is not only the top management prize at Huawei, but also a reflection of a clear acknowledgement by Huawei leadership that Huawei's success is largely attributed to its having learned from the West, especially the United States. Huawei's management, from its systems to its processes, is deeply branded with the influence of modern US organizations. A telling example of this is that the most frequently used management-related terms at Huawei are all English abbreviations— typical IBM style.

Who are the Whiz Kids? To some extent, this phrase is synonymous with modern US management. After World War II, 10 veterans from the US Air Force joined the Ford Motor Company. They had all gone through the same rigorous screening; most had attended Harvard Business School and had received military training; and they all had a similar style of speaking. Together, they promoted and formed a new management cult in the 1950s. The starting point of this new "cult" was to base everything on numbers and facts, emphasize control, efficiency, and order, and respect the corporate system as if it were a religion.

When the Whiz Kids joined Ford Motor, chaos dominated the company. It didn't have a basic organizational configuration and no data could be used as the basis for decision making. The concept of "planning" was unheard of, let alone budgeting. Communication channels were blocked up and corruption prevailed. To illustrate, a director in charge of 7,000 people didn't know whom he was supposed to report to. For 44 years since its establishment, Ford Motor had never audited its accounts. And in 1946, after suffering 13 years of consecutive losses, it had been in the red for the longest time of all corporations in US history.

Fortune magazine described the state of most US companies in the 1950s as being full of ambitious magnates, adventurers, and speculators, as well as illiterate foremen; laudatory of individual heroes and barbaric growth; reliant on personal perseverance and determination, rather than on professional business managers and expert talent; governed by personal intent, not by systems and processes.

After nine years of sweeping reform, the changes at Ford Motor were apparent. They had an organizational chart and they had defined the boundaries of delegated and central authority. Decisions were made based on facts and data. Cost control was significantly tightened and product quality was improved. Performance management was highly valued across the company. In less than a decade, the Whiz Kids had transformed the once chaotic organization into the third largest company in the United States.

Invisible changes were also exerting their influence on the organization's culture—order was more valued than disorder, data was more important than personal whim, tools were more effective than arbitrary ideas, rational decisions were better than impulsive ones, and wisdom and hard work won out over deceit and opportunism.

Business is truly the spirit of the American people. The Whiz Kids whipped up a powerful whirlwind of change in US business, which then swept through the army and the government. The whirlwind culminated when Robert McNamara, one of the Whiz Kids, was invited by President

Kennedy to serve as Secretary of Defense, and to use the same "scalpel" of facts and data to transform the US military forces.

Needless to say, IBM represents classic US-style management: focusing on the collection and analysis of facts, respecting rules and procedures, emphasizing the function and strength of organization, extracting efficiency from figures, and predicting trends and changes through data analysis. This is exactly what was missing at Huawei and the vast majority of Chinese companies in their formative years.

Some have questioned why Huawei chose IBM as its consultant, a company known for its stable and conservative culture, when Huawei had also visited Cisco and other US companies in 1997. Ren Zhengfei argued, "Cisco is too radical in its management, and Chinese people's minds are supple enough as is." The prudent, even conservative management of IBM might have been a better fit to help mitigate the chaos and vitality of Huawei in its early days.

In its formative years, Huawei's R&D activities left a lot to chance. The success of the C&C08 was attributed to the pronunciation of the name, as "C08" sounds like "prosperous all the way" in Chinese; likewise, the failure of the EAST8000 was blamed on sounding like "easy to die" in Chinese. For an organization consisting primarily of intellectuals with engineering backgrounds, irrational belief in numbers and feng shui was not at all uncommon in Huawei during its first 10 years. One story has it that after three consecutive nights of no progress in repairing a machine for a customer, they burned incense in offering to the gods and the next time they started up the machine, it was up and running again.

There was another story from an overseas rep office. During a project that the team had made no progress on for a long time, someone pointed out that the painting on the wall behind the regional president's desk was destroying the feng shui of the office. Shortly after the painting was replaced, they signed a contract with the customer.

What is the reason behind such superstition? The lack of operating processes and weak organizational structure naturally led to irrationality.

In the introduction to this book, I began by talking about the numerological superstitions of southerners in China. Behind that explanation was an implied question: How should we define Huawei's success? Is it inevitable or completely by chance? Is it predestined or driven by a certain logic? Or perhaps a little bit of both?

In its infancy, supply-chain management at Huawei was vastly inadequate. As some described, "the shoes were never made big enough for the feet, which were growing fast." As a matter of fact, the market grew so fast that they often couldn't deliver on orders after contracts were signed. As the joke went, they were "clubbing every night (to win contracts) and firefighting every day (to fill the orders)." Someone also quipped: "We couldn't have solved our delivery problems even if we had found a project manager from the moon."

For a long time, there was no basic structure for business planning and market forecasting in Huawei. Their approach was likened to the process of making dough: "Add a little more flour if there's too much water, and add a little more water if there's too much flour." Supply shortage and inaccurate deliveries occurred all too frequently, so they even set up "Delivering the Correct Equipment Team" to turn things around.

Prior to 2007, Huawei's financial system was pretty much an old book-keeper, fulfilling functions like collection, payment, and reimbursement. Some even went further to argue that, for many years, no one in Huawei could say for sure how much profit or loss had been made for a given year; there were only rough figures. Project management and detailed business analysis were certainly out of the question.

Huawei's early predicament was characterized by a coexistence of chaos and vitality and the company's development was largely shaped by individual heroes, a combination that made it extremely difficult for Huawei to expand and succeed in the international market. For a company to go global, it has to have standardized management and complete systems and processes that are fully aligned with international—particularly Western—rules. This was a giant wall to climb over and the ladder was handed over to Huawei by IBM and other Western companies.

What did these Whiz Kids bring to Huawei? As the company's rotating CEO Guo Ping summarized, "Huawei's approach to management has evolved from qualitative to quantitative, or from 'literature' to 'mathematics,' so to speak. Now we can manage our business in real time based on data, facts, and rational analysis."

Xu Zhijun, another rotating CEO, who ran the company's R&D function for many years, said,

> In the past, when we were managing an R&D team of 3,000 people, it was fevered and hectic.[2] But now I have a team of 70,000 and I can manage it pretty well. Even 140,000 won't be a challenge. You only need to work with a handful of people on each product line, because they have the same processes and management system. As long as you get the leadership team right, you're done.

Through the ISC transformation, Huawei has established an integrated, global supply network that covers every major region of the world, bringing the company closer to its customers and effectively supporting the company's global expansion. Through continuous reform programs, services and delivery have become a strong competitive advantage and an important means of enhancing profitability. The IFS (integrated financial services) reform has helped the company build up a global financial management system, which has closely integrated the business with its finance function, driving fundamental progress in cash flow acceleration, accuracy of revenue recognition, visibility into project profits and losses, and business risk control. At the same time, with the help of Hay Group and other Western consulting firms, Huawei has established a comprehensive human resources-management system over the past 27 years, which covers recruitment, appointment, retention, training, and management of its people. This system has played a crucial role as Huawei evolves into a leading global company.

So how should we define Huawei nowadays? When exchanging some ideas with authoritative experts from Western countries, a top executive at Huawei said, "The only difference between us and you guys is ethnic background. That's it."

---

[2] Xu Zhijun uses an extremely colorful expression here, which literally means "we spent all day fighting malaria." He's not being literal, of course, just describing how chaotic and feverish R&D operations were back in Huawei's formative years.

British Telecom, or BT, once the world's largest telecommunications company, was practically decimated during the dot-com crash of 2001. Cost control became the top priority when it planned to upgrade and transform the United Kingdom's entire telecommunications network. BT's board turned their eyes to Asia, especially China. After visiting several Japanese and Chinese companies, they finally decided to identify Huawei as a primary candidate.

BT is well-known for being a difficult-to-please customer in the telecom industry. It had a demanding certification system, comprising over 100 criteria across 12 categories (including what Huawei people called "human rights certification" to ensure the compliance of, among other items, the working environment and overtime pay). For a Chinese company with a history of just over 10 years, the certification program was undoubtedly a huge challenge.

BT left no stone unturned, but in the end, Huawei managed to survive: It passed its certification and was entered into BT's supplier shortlist. The certification experts at BT were particularly impressed with what the IPD reform had brought to Huawei—a reliable, sustainable, and testable product development process. It gave the British "a familiar and comfortable air."

"Forgetting the past means betrayal," Ren Zhengfei remarked at the Whiz Kids award ceremony. Obviously, this expression is rich in meaning and can be interpreted in many ways.

Following Established Rules: Conservatism Is a Good Thing

Corporate Crew Cuts: Going from Shenanigans to Soldiers

Most companies die in their first one to two years of life and most of them die of hunger, not disease. They are like pirates who failed to find treasure and are forced to either sail to shore or drown in icy-cold waters. Others survive for three or four years, even longer in some cases. Their business might prosper and their teams might expand, but the outlaw spirit, hero culture, or cliquishness that developed early on will only serve to hinder further growth.

During this period, entrepreneurs instinctively pursue institutional reform to put an end to the old outlaw spirit and create new order. We call this type of reform "fundamental reform" and no matter whether it's a conscious decision or a decision that is subconsciously forced on the organization's leaders by circumstance, it's a process whereby the enterprise goes from a state of chaos to a state of stability at an underlying or superficial level.

In general, when a company grows to be unbearably chaotic, many founders—particularly those in China—tend to launch strident, top-down institutional reforms to eradicate the company's ties to primitive passion (pejoratively, the outlaw spirit). They do this in the hopes that they'll wake up and everything will be different the next day. But typically, their intentions often produce the opposite results. Following radical reform or shock therapy, a company usually faces one of three possible outcomes. First, the shock of a massive earthquake or colossal typhoon and then open confrontation of opposing forces amidst the rubble—this is one possible outcome. Another is that silence infiltrates the company, creating an atmosphere that is depressed, passive, pessimistic, cautious, lazy, and perfunctory. In the worst-case scenario: mutiny, and all the folks who are worth their salt jump ship.

I randomly interviewed over a dozen business leaders and employees on their understanding of the word "institution" and the answer was essentially the same: Institutions mean control, restriction, prohibition—one big "that's not allowed." Imagine a horse galloping at full speed and then suddenly slamming into a solid steel wall. Man and horse alike would be flat on their backs, maimed if not gone for good.

When they smack into the wall,[3] many adaptable entrepreneurs take this as an opportunity to build out their systems or engage in a little

---

[3] Here, the author refers to the phrase "run into the south wall," which is an interesting reference to historical architecture in China. In the traditional houses of affluent families, the

institutional patchwork. Wherever there's a leak, they slap a system on it, and wherever there's a problem, they prop it up with a rule. These systems and rules are often shelved until a new problem arises, at which point the patchwork begins again. As mentioned previously, the vicious cycle of the "add a little more flour" approach leads organizations to rot from the inside out.

Yet still, China has a handful of entrepreneurs who, like a mouse chewing through the baseboards of the market, through 10–20 years of trial and error, opportunism, and resilience, have built up noteworthy—even world-class—companies. After all, in a country that has virtually no business philosophy or experience, business people can only draw inspiration on corporate governance from China's political culture, and, as Deng Xiaoping put it, "feel around for stones while crossing the river." Over time, through one jolt after another, they have slowly, gradually managed to understand and incorporate tried-and-true practices, applying them where necessary to develop a unique set of cultures and systems. Lenovo, Haier, Vanke, and Midea are all perfect examples of this. Although what they've become today differs greatly from what they had originally planned for, these companies are, after all, the pride and glory of Chinese business history.

Their success is, in part, a result of the opportunities created against the backdrop of China's reform and opening up, but no one can overlook the wisdom, resilience, vision, and willpower of these company founders when they went through their periods of fundamental reform. Lenovo founder Liu Chuanzhi, Haier founder Zhang Ruimin, and Vanke founder Wang Shi are not radical reformers or inventors of new management philosophies. Instead, they repeated the same ideas over and over again, tirelessly instilling them in their employees until these ideas evolved into each company's own set of core values. When it came to institutional

---

main entrance into the household or courtyard was in the south. A few feet in front of this main entrance, they would construct a screen wall (also called a spirit screen) made of wood or stone that served two purposes: It blocked the line of sight directly into the house (for privacy) and also kept out evil spirits (which in ancient Chinese cosmology couldn't turn around corners). When you leave the main door of the house, you would have to immediately turn left or right to go around the screen wall. So in China, if you say someone "ran into the south wall and didn't look back," it means that they are stubborn and refuse to listen to other people's opinions.

reform, they excelled at compromise and buying-out strategies: They compensated people who had vested interests for their obedience and willingness to submit to the rules when their power was restricted or stripped away. Of course, they would also employ the corporate version of capital punishment to deter ill intentions and protect authority. They would use sudden, heavy-handed measures at critical moments in critical reforms, but in general they planned well ahead.

Successful entrepreneurs like this are like turtles, calm and resilient, but they are also chameleons that can change colors at will to protect their companies against harm.

Wang Shi, the CEO of Vanke, once said that Ren Zhengfei is a cunning fox. At the time, he perhaps didn't realize that he was actually describing himself—painting a picture of that entire generation of entrepreneurs.

Fundamental reform is inevitable for any company (not including SOEs and other companies that depend on resource monopoly—the arguments in this book only apply to market-oriented private enterprises). The key is to build a system of rules and procedures that successfully removes it from the backwoods mentality. This type of reform is like getting a corporate crew cut. Powered by the establishment of systems, it trims away the wildness and the lawlessness that grew up out of the company's primitive period of hunting and gathering, leaving behind the tufts of chaos and scruffy locks of disorder to produce a dapper and smooth, modernly governed corporation. The methods and tools of reform are important (more discussion of this to follow), but the nature of the new system is key.

A bad system is prohibitive, much in the same way a cage is for birds or a pond is for fish. Under such a system, majestic eagles are reduced to canaries and august whales are no better than dumb, mud-sucking fish. If outlaws are tamed into a bunch of abstemious and withdrawn holy men or polished into perfect gentleman, how can a company—whose primary goal is to seek out profit—possibly stay in business?

A good system is like a racecourse with clear boundaries or a river with a sturdy dam. It tolerates eccentricity while taming primitive passion, inspires vitality while suppressing apathy. In such a system, bridles are put on horses, outlaws are transformed into soldiers, and guerrilla troops become regular units.

The difference between a bad system and a good system, I strongly believe, deserves the attention of Chinese companies and their leaders that are moving towards modernization.

This reminds me of the fireplaces at Huawei. At the company's headquarters in Shenzhen, which is a subtropical city, a European-style fireplace adorns certain reception facilities and conference rooms. When the city's so-called "winter" arrives, even though it's not really cold, these fireplaces must be lit when Ren Zhengfei receives a guest (in many instances, he lights them himself, adding lots of firewood in the process). The room is always warm and cozy, lightly fragrant with the scent of birchwood, the smoke rising up through the chimney, the fire dancing away on the hearth, and the words flowing confidently from the mouth of Ren Zhengfei.

Why does he love these fireplaces? If we look at it from a psychological perspective, what does this preference mean?

After the publication of the first edition of this book, Ren Zhengfei likened it to a scene from the TV series *Outlaws of the Marsh*: a crowd shouting in unison around the campfire: "To the fire! To the fire! To the fire!"

Fire is a symbol of the organization's passion and it also brings about cohesiveness—but the burning fire must be contained within the fireplace.

Nothing but Magic Can Enact Speedy Reform

Fundamental reform or institutional development requires that companies bid farewell to their formative years of hunting and gathering. In most cases this reform is radical, wide-ranging, and meets with strong resistance. Leaders and reformers require vision and insight, the courage to take risks, and patience. Institutional development does not happen overnight and there is no such thing as a perfect institutional design. A good system is formed through ongoing revision and improvement: As long as the overall framework is correct, the company must stride forward, steady and true.

As for reform at Huawei, the company staunchly opposes the following seven things:

1. Perfectionism

2. Hairsplitting

3. Blind innovation

4. Partial optimization that doesn't improve overall efficiency

5. Managers leading the reform without looking at the big picture

6. Employees without practical experience participating in the reform process

7. The application of any unproven processes

These seven items, the so-called Seven Prohibitions, represent the core elements of Huawei's reform philosophy, the guidelines of its reform. As the saying goes, "When you've got the right key, all the pins in the lock fall into their proper place."

Ren Zhengfei said at a meeting in 2009:

> For over 20 years since its establishment, Huawei has never stopped transforming; However, we don't pursue dramatic change, because this comes at a hefty price. We have moved ahead with gradual, evolutionary changes, changes that perhaps no one has noticed. Transformation can't be dramatic. It's not aimed at producing heroes that command great forces at their beck and call. This would ruin the company. We can't all pay the price for one great person's success.[4]

The time frame set by IBM for the IPD transformation was nine months, but it took them four years. The IFS transformation was expected to finish in two and a half years, but it ended up taking eight years. Huawei's rotating CEO Guo Ping said, "What we want is the best results, not to

---

[4] Here Ren is referring to a line written by a poet named Cao Song from the late Tang Dynasty. A prosaic rendition of the line is as follows: "Please don't talk about the issue of rank (earned through war) / A general's success is built on thousands upon thousands of bones."

put on a show. You can't just set off a firecracker and call it a deal. Just going through the motions isn't what we need."

Nothing but magic can enact speedy reform, but IBM isn't a magician. Neither is Ren Zhengfei. All those magical acts of reform out there, once exposed, are nothing but smoke and mirrors.

Earlier in 2000, Huawei's leaders made a clear declaration: "Organizational adjustment and development should be gradual and evolutionary, not revolutionary. There shouldn't be any abrupt changes. Transformation must expand and contract at a controlled and appropriate tempo."

In this statement, transformation was defined as evolutionary, not revolutionary. Why? And what's the difference? Revolution is destructive, a complete change in dynasty where you plow everything to the ground and start over again. Revolution is very costly, usually shocking, and damaging. China's history is filled with violent revolutions, with unrest and turmoil lasting longer than peace and stability. Consequently, China lacks continuity in its institutional systems despite a rich cultural heritage. One dynasty overturns the old one, then sets up a new system. Even the buildings themselves are put to the torch. An article in *The Times* ("Chuck Your Chintz to China. It's Valued There," February 6, 2012) discussed this phenomenon:

> In China the new is destroying the old, and concrete structures are driving up a ruthless and destructive tsunami. What psychological price the nation has to pay when concrete structures bury its past? Perhaps the people now are much richer and healthier than their ancestors who had tilled the land with sweat. But if the collective memory is erased, do they still know who they are?

This is the general disposition of the nation and it has surely penetrated the corporate culture of businesses in China. Entrepreneurs should be keenly aware of this.

Business, however, is not like art. Artists need to defy and renew themselves, but a business organization, like a nation, needs heritage and continuity. A Huawei executive once said, "A company is a constructive organization. You can't just go about making radical transformations and innovating your management. Change is necessary for growth, but companies must change with caution."

Several years ago, Ren Zhengfei told the story of Xiao He and Cao Can. Xiao He was the first prime minister of the Han Dynasty, and Cao Can took over his office after Xiao retired: "When Cao Can first took office, he didn't repeal any rule that Xiao He had established, which helped ensure the continuity of governance. Huawei would like its managers to act like Cao Can. We can't be preoccupied with innovation all day long." In fact, "innovation" is rarely mentioned in Huawei's documents and is not considered part of its core competitiveness, especially in terms of corporate management. Ren said, "All innovation has a price, and it will harm the company if the price outweighs the returns. Over the years the company has accumulated a vast amount of management procedures, so any arbitrary innovation would be a waste of our past investment."

Conservatism is the antonym of innovation. According to Italian painter Jannis Kounellis, in Italian, the word "conservatism" means the effort to conserve everything great from the past. To be conservative is not at all a decadent thing. Even if we launch a revolution, we are doing so to return to our past glory and then continue the journey from there. This is why the Renaissance, the Italian revolution against the Dark Ages, refers to the rebirth of classical tradition; it didn't sever its link to history.

In human history, progressive and conservative, radical and reactionary, Left and Right—they have all meant positive or negative things depending on the context of time and space. But it is also very likely that progressivism and conservatism offset, interchange, coexist with, and check one another, and that such interactions constitute the main theme of human history. Perhaps the history of a company is no exception.

Portrayed as an innovative company by the media, Huawei has never agreed that innovation in corporate management is part of the spirit of the organization. Instead, Ren Zhengfei argues, "What's wrong with conservatism? To be conservative is to maintain continuity and stability. Is it really necessary to break every single tradition to be progressive? I don't think so." In 2013, China experienced a sudden Internet craze and some people at Huawei also felt the itch to get on board. In response, Ren Zhengfei was vocal about embracing the "Tortoise Spirit," warning the company not to chase after the winds of fashion but rather to stay focused on the core business while keeping an open mind to all the mental tools out there (including the Internet) that might be beneficial to Huawei's development.

In 1999, when IBM's consultants began implementing institutional reform at Huawei, Ren pushed forward the aforementioned "cut the feet to fit the shoes" initiative and took definitive measures to eliminate resistance to the reform. At the same time, he remained sober and insisted that

> we must develop our management system by improving existing processes. We have to continue optimizing non-value-added processes and value-added processes alike—continue to evolve them and never stop, getting as infinitely close to a reasonable structure as we possibly can. Our long-term guiding principle is to offer big rewards for small improvements.

For many years, there have been frequent news reports that praise and commend forward-thinking big shots who propose or make significant changes to management in their respective companies. In this, Huawei remains an exception that doesn't encourage recommendations for sweeping change. Why? Management is a systematic project where an abrupt change might lead to widespread upheaval in the company. Conversely, minor changes or improvements are helpful for optimizing the system and are therefore more constructive and practical.

To get infinitely close to a reasonable structure is the ideal tempo of reform in the eyes of Huawei's leaders. As per Ren Zhengfei, "reform is an ongoing process. We shouldn't be too radical. If we make 0.1 percent progress every year, we will achieve 10 percent in 100 years. If we can sustain this type of progress, our long-term improvement will be incredible."

Perhaps 100 years is too long. By then, Ren Zhengfei and his contemporaries will have left the stage. But Ren isn't interested in just seizing the day. If the company attempts to do so, nothing would last and the company would have to start all over again. This type of tragic decision is common throughout Chinese history.

Li Zehou, a famous Chinese philosopher and scholar, argued—quite controversially—that China must bid farewell to revolution. Li explained:

> China spent three-quarters of the 20th century revolting, and revolution became a holy slogan as if revolution is a good thing. In fact, revolution is not sacred. We are opposed to revolution or the act of deifying revolution. Without it, perhaps society would be better. The French Revolution lasted many years, but caused France to lag behind the UK, which had developed a stable regime and adaptable political system. [...] The 1912 Xin Hai Revolution delayed China's process of modernization, proving in some way that evolutionary reform is better than abrupt revolution.

I don't know if Ren Zhengfei has read Li Zehou's argument against revolution, but I strongly believe that Ren agreed with him when he said,

> Liang Qichao, a reformist during the late Qing Dynasty, talked extensively about revolution because he actually wanted reform. To me, bidding farewell to revolution is to allow gradual reform. Some say that revolution is easier than evolution. I partially agree. I believe revolution is very tough, but evolutionary reform is even tougher.

Ren Zhengfei was born in the Year of the Monkey, which generally implies a restless and energetic character. It is of great surprise therefore that Ren has remained calm and patient while building Huawei from a small dinghy into a mid-sized aircraft carrier. His reform has been solid, gradual, and perhaps conservative, which is antithetical to his Chinese zodiac sign and personal disposition. No doubt it has taken great will to restrain and deny himself, and in effect overcome himself.

The bigger challenge is overcoming the restlessness and monkey-like enthusiasm of the team. And then there's the question of whether or not evolutionary reform on its own is enough. Over time, mild reform might not be enough to clear away all of the organizational dirt that has accumulated; if the organization were to become clumsy and slow, would radical revolution be necessary, even if it harms the organization? Where's the critical point between progressive and radical change?

The Eight Major Symptoms of a Company That Needs Reform

A fatal trap often lurks around organizations in their early stages: their dependence on heroes. Whoever can bring in products, contracts, or revenue to help ensure the company's survival is deemed a hero. As a result, the people in these roles will get higher salaries, more bonuses, better stock options—and more power. Yet at the same time, when they reach a certain point, heroes will begin to rest on their laurels, threatening the collective will of the organization. This forces the leaders to make a choice: to transform the company into a highly institutionalized organization.

But the problem is, when is the best time for institutional development? Change always comes as the last resort. Technically, Huawei's fundamental reform—or institutional transformation—started with the Huawei Charter in 1996. Was it possible for Huawei to begin reform earlier? The answer is no. Ambition and desire are two sides of the same coin. If there is no "infancy stage" of bustle and chaos, of vitality and turbulence for a start-up, if there is a complete set of standardized and orderly systems at its establishment, the odds are that such a company will never grow up and will remain a premature baby forever. The early position of a business is largely shaped by laissez-faire, or even barbaric growth. Fearlessness and adventurousness, to the extent that the law permits, are the primitive genes necessary for any start-up. Prematurely changing systems and processes will stifle heroes and hinder the rapid growth of an organization.

Problems and desperation are the driving forces and catalysts for change. And leaders, as it happens, are born to solve problems. Heroism that has brought about prosperity might also breed mutation—from aspiration to greed. This is the time when daggers of reform must be directed towards the people or things that the leaders had once appreciated and encouraged. This is a peculiar situation, but it's an inevitable choice that decisive leaders have to make over and over in their lives.

In general, there are eight major symptoms to watch out for that might be cause for major organizational change:

1. Do deviations or contradictions occur in corporate values when they are presented, communicated, and implemented? Does

"customer centricity" turn out to be "leader centricity"? Or do people lean towards a combination of both customers and leaders? To what extent are salary increases, bonuses, stocks, and promotions issued based on performance and contribution?

2. Is employee engagement a major problem in many parts of the organization? Symptoms to watch out for include general dissatisfaction with compensation, overwhelming complaints about career opportunities, widespread suspicion of inequitable treatment, and collective laziness.

3. During the early stages of rapid growth (at about 5–8 years old, the primitive hunting and gathering period), are there any signs of a split in the organization's power structure? Examples include major or minor factions, business executives with too much power, confrontation between executives, or executives who use their factions to threaten the company with unreasonable demands?

4. As the organization's systems and processes become more established, are there any constraints in the power hierarchy, both vertically and horizontally? Do they result in a massive decrease in decision-making efficiency and in anxiety within the management team? For example, the proactivity and creativity of leaders and managers appears to have been suppressed, followed by large-scale organizational collapse as people resign in droves.

5. Has technological and product innovation stagnated for an extended period of time or has it suffered several major failures in a row?

6. Do any major issues stand in the way of business growth?

7. Have there been any drastic changes in the external environment, like a reshuffling of the industry landscape, sector-wide bottlenecks in growth, or challenges from other industries?

8. Is there a growing number of complaints from customers?

These eight symptoms rarely appear each on their own. In most cases, several appear at the same time, and reform therefore becomes a pressing imperative.

Reform is the soul of progress for a business. At every step forward, companies have to face countless unknown variables and risks. The mission of entrepreneurs is to seek certainty out of uncertainty, to navigate the unknown, find the right direction, potential points of breakthrough, and identify the right tactics to achieve victory. In that sense, entrepreneurs are mostly adventurers: They are inclined to gamble, they can handle stress well, and they are galvanized by anything new and different. Even for those century-old companies, while those at the helm keep changing, all the successful leaders have led their companies forward by turning their back on tradition. In a sense, the secret for any business to sustain its development is to innovate and to reform. And yet innovation brings destruction and reform is often accompanied by turbulence. So balancing innovation with order, pacing reform in a measured way—this is the tricky part of corporate management.

Huawei has been successful in all of its major reforms. But an important thing to note—and this is something that people often overlook—is Huawei's conservative, incremental, well-paced, and forward-thinking approach to change. Almost every reform was launched right as these symptoms just began popping up, when the business was still in good shape. The company would start with a little "drizzling" to loosen up the soil a bit, then move forward with the unrushed steps of mild reform. Some might wonder how Huawei's leaders have been so precise in orchestrating their reform programs. What's the trick?

The trick is customer satisfaction! It's where all of Huawei's reform initiatives begin and end. In other words, the eighth symptom is the fundamental one, the pulse that indicates every single step the organization has to take for the prescribed reform to be targeted and effective.[5] This is the ultimate source of judgment at Huawei and it is also the sole criterion for evaluating any person or any department within the company. Looking at it from this pragmatic perspective, the existence of some seemingly contradictory reform programs—or the "negation of prior negations"—should come as no surprise.

---

[5] This is a reference to Traditional Chinese Medicine, in which the physician feels the patient's wrist and diagnoses any problems by feeling for irregularities in the patient's pulse.

On Change (II): Purpose above All

Chapter 8

Culture: The Mother of Systems

The Huawei Charter: Their Very First Top-level Design

In 1995, Huawei welcomed several young scholars from Renmin University of China, including Peng Jianfeng, Bao Zheng, and Wu Chunbo. They had all studied overseas in the United States or Japan. After returning to China, they translated a large number of US and Japanese management books and began to promote these works among domestic companies. They were the first wave of "Promethean" management scholars after China's reform and opening up, and Ren Zhengfei invited them to design the compensation packages for Huawei's sales force. Prior to this, Huawei didn't really have any decent systems.

In early 1996, on a whim, Ren Zhengfei decided to invite these young specialists to develop a "Huawei Charter." Huang Weiwei, another scholar who had returned from the United States, came to help and ended up forging a bond with Huawei that has lasted ever since. During the 20 years after Peng Jianfeng and some others ended their partnership with Huawei, Huang Weiwei and Wu Chunbo kept serving Huawei as management advisors, making close observations and researching the company.

In his praise of Huang Weiwei, Ren Zhengfei said, "He has a very firm academic foundation. How many people nowadays read those massive original works from the West? Mr. Huang devours one volume after another." Huang Weiwei was hired as Huawei's chief management scientist in May 2014, the third year after he retired from Renmin University.

The two-page framework of the Huawei Charter was drafted by Huang Weiwei and Bao Zheng to answer three questions: Why has Huawei succeeded? Can its past success lead to greater success in the future? What's needed to achieve greater success? There are several modules within this framework: strategy, organization, human resources, and control—a classic system of management in the West, but not many people in Chinese companies had a grasp on these concepts back then.

The so-called "Six Gentlemen from Renmin University," Peng Jianfeng, Huang Weiwei, Bao Zheng, Yang Du, Bo Wuchun, and Sun Jianmin, helped infuse Western blood into Huawei's Eastern veins. This move, though quite common nowadays, was undoubtedly significant over 20 years ago, when the company was still feeling around for stones while crossing the river.

The main section of the Huawei Charter focuses on the value of helping customers realize their dreams. This section set the fundamental direction of Huawei's values for the 20 years to follow. The company had adopted a technology-driven approach in its early years, but was technology the most important thing or customers? In 1996, this question of technology versus customers was the object of intense internal debate. However, the charter clearly stated in its opening chapter that Huawei's position is to "realize customers' dreams," so in effect, it settled the debate in the form of "company law."

Less than a year later, Huawei chose IBM's management consultancy services over those of other American companies; the key consideration was IBM's market-driven philosophy. "This impressed the boss," Sun Yafang recalled. Later, the end-to-end (starting from the customer and ending with the customer) value proposition that IBM promoted throughout the IPD transformation not only reshaped Huawei's R&D system, but the company's entire organizational structure.

Essentially, the fundamental ideas in the Huawei Charter and IBM's business philosophy were cut from the same cloth.

From Huang Weiwei's point of view, the Huawei Charter is a systematic summary and elaboration of Ren Zhengfei's management philosophy. For example, Ren believes that "resources will be exhausted, only culture will last forever. Huawei has no natural resources to rely on, but we have big oil wells, big forests, and large coal mines to be explored in our heads." This actually defines Huawei as a company that centers its operations around people. Human resources are the greatest assets of the company. With this belief, Huawei is defined as a value-creation and -distribution system comprised of financial capital and labor capital, which includes intellectual output and entrepreneurial contributions. This is also where the core value of inspiring dedication[1] came from in 2009.

In the first draft of the Huawei Charter, Ren Zhengfei added the line, "In order to make Huawei a world-class equipment supplier, we will never enter the service industry, so that market pressure as a result of being non-dependent will keep our internal mechanisms in a permanently active state." This is Ren Zhengfei's basic business principle and it has forced Huawei to take the only road possible: relentless focus without any diversification. Ren's remarks were a source of great controversy in the company, but against all opposition, he kept emphasizing that "not a single word can be changed."

According to the Huawei Charter, "10% of annual sales must be spent on R&D, and this investment will be increased when necessary and required." In fact, the company had been doing this already, but it was made official in the charter.

Ren Zhengfei also made several additions to the second draft of the charter: "no losses for the selfless"[2] and "no compromise for past heroes

---

[1] "Inspiring dedication" is not as simple as it first appears. It is a robust and well-established system in Huawei that involves monetary and non-monetary incentives to inspire people to work harder. By working harder, employees create more value for customers, value which is then distributed back to the employee in proportion to their contribution, thereby motivating them further.

[2] Translator's note: The Chinese phrase here is literally "Don't let Lei Feng suffer losses." Lei Feng is the name of a legendary soldier and cultural icon in contemporary China. He was

that lag behind." These left a deep impression on Huang Weiwei. He could practically smell the threat of blood in the "no compromise" statement—it pinpoints a major pitfall for start-ups once they reach a certain stage of development and Huawei had defused it by setting an iron law. This addition also signaled the fact that Huawei was starting to develop in a more contractual way, with less emphasis on relationships between people.

They revised the Huawei Charter eight times, discussion and debate lasting for as long as two and a half years. The charter combined the essence of Western management, refined versions of successful practices that Huawei had gleaned over the past 10 years, and the intellectual spark of quite a few young managers in the company. More importantly, it cut to the heart of Ren Zhengfei: the founder's enormous ideals, his nationalistic convictions, as well as his complete set of management ideas based on deep insight into human nature.

Objectively, the Huawei Charter was important because it unified ideas and helped drive consensus throughout the company. At that time, Huawei's rapid development had brought it to the brink of a major potential crisis; different philosophies abounded, filling the company with vitality and also flooding it with doubt. While the black cats and the white cats sure did catch a lot of mice,[3] ideological and organizational unrest began to surface due to the free play of different factions within the company. If the situation had gone on without any rules or restraints, it would have led to a series of problems.

The Huawei Charter is the first "top-level design" in Huawei's history. Although the draft designers were mostly scholars who had studied in the United States or Japan, its content, structure, narrative pattern, and the way it was communicated still had strong Chinese characteristics—ideological mobilization was more important than the design of the

---

a prominent figure in positive propaganda during and after the 1960s, representing a model citizen: selfless, devoted, and modest. Here, Ren uses the personification of Lei Feng to refer to Huawei employees who demonstrate these same characteristics.

[3] If you will recall the Introduction to this book, this is a reference to a famous quote from Deng Xiaoping: "It doesn't matter whether it's a white cat or a black cat; a cat that catches mice is a good cat," which means that differences in ideology are unimportant as long as people contribute to China's development.

system. In a sense, it served as a prelude—or a way of getting rid of the salt in the soil, so to speak—for subsequent institutional reform.

Of course, the Huawei Charter is not a set of laws, per se, because the compliance of Huawei's shareholders, managers, and employees is not mandatory. It is more akin to a document of guiding principles for that particular period in Huawei's history. For example, the company has never implemented the business-unit structure that was written into the charter. After all, the document is intended as a mild philosophical constraint, written in the borrowed name of law.

Soldiers Are Meant to Be Used by the General, Not Owned

In 1996, at around the same time the Huawei Charter was being drafted and discussed, Huawei's sales department underwent a major shake-up that lasted about a month. All of the people in leading management positions, both in the head office as well as local cities in China, had to submit two reports: the first, a performance summary of the past year with a plan for the coming year, and the second, a resignation letter. The company would either approve the work report or accept the resignation based on each individual's performance, their growth potential, and the company's market development needs.

This was a shocking move. Several years later, an executive at Motorola China recalled the event, saying that only Huawei would ever dare do such a thing—and succeed, no less. If someone did the same thing in most Chinese or US companies, who can imagine what would happen? The turnover rate was already high in the sales team: It took many years and a lot of resources to develop a regional manager. If everyone left, it would be a huge loss, not to mention the customer relationships that they might take with them. No one had wanted to lose a single sales manager, but the company had asked all of them, including the president of the sales department, to resign. Only Ren Zhengfei had the guts to do such a thing. Something like this might have been common in the military, but it was taboo in the business world.

However, this event didn't cause the least bit of stir within the company. Quite the opposite: The massive resignation evolved into a rallying call. Sun Yafang, then president of Huawei, delivered a collective resignation speech on behalf of the sales department. She said that it was a great act of defiance to throw in the face of bureaucracy and hierarchy and that it proved that Huawei's people could truly move up or down the ladder depending on their contribution. After a team representative read his resignation letter out loud on stage, a number of people from the sales department were inspired to stand up and speak:

> For the sake of the company, I am willing, without a single complaint, to sacrifice my own personal interests.

> Huawei's culture is characterized by solidarity and growth. As a Huawei person, I am willing to be a stepping stone for the company's progress.

> For the sake of Huawei's lasting success, I shall not drag the company down for individual reasons.

Employees from other departments also spoke with approval:

> Why is Huawei so unique? Because its employees are united as a family, and each is willing to sacrifice his individual interests for the sake of the organization.

> I've been working here for many years, but I didn't understand the real meaning of sacrifice until today.

Mao Shengjiang, the acting sales president, spoke to the heart of the matter. He said:

> Anyone stepping down from a prestigious position can't help but have mixed feelings; it will take time to recover. I'd be lying if I said I didn't care. No one ever tries to make trouble for themselves, willingly. Anyone who says otherwise is just trying to make themselves look good. It's not real either.

> When I stepped down as the acting president for sales, I had not expected to learn as much as I know right now. People around me keep asking me if I care about all this. Surely I do. I care about the company's growth and success. I care about the youth,

passion, and sweat that we've given to the company. I care about what I can continue to do for Huawei; I care about the confidence that I derive from the job; and I care about whether or not I can fulfill the wishes of my comrades-in-arms. Prestige, position, all those other vain things—I don't care about those at all.…

The sales department had been around for five years and during that time, they had made great contributions, distinguishing themselves as a heroic pack of lions within the company. So in its second round of growth, why had the company decided to disband the team and build a new one? Decision makers at Huawei explained:

> First, our competitors are so powerful that we haven't seen the true extent of what they can do. They're like elephants, and we're like a mouse. If a mouse stands completely still, it would be trampled to death by the elephant. But if the mouse is nimble and moves around quickly, it can climb up the elephant's back, or even up its nose. And if the mouse can consistently dodge the elephant's feet, the elephant will eventually lose its cool. The idea is that we can't be conservative, rigid, or static; our operational system and organizational structure must be flexible.

Indeed, this was perhaps the only way out for a private company that began with only CNY20,000. Since day one, Huawei has faced a number of giant rivals right on its own doorstep and it had very few advantages. The only way it could possibly win was for its people to work as fiercely as lions, advance as rashly as wolves, be more nimble than elephants, and move about as deftly as mice.

This is the reality that Huawei faced all throughout its first nine years. And that reality is what forced the company to shake up its sales department.

The second reason, they explained, was that "soldiers are meant to be used by the general, not owned by him." Some people have likened Ren Zhengfei's business philosophy to Zeng Guofan's[4] approach in managing troops, believing that the two bear a striking resemblance. This is a misconception. There is no evidence that Ren Zhengfei has learned anything

---

[4] Zeng Guofan was a statesmen and military general who raised a standing army, called the Xiang Army, to suppress the Taiping Rebellion, a civil war that broke out during the late Qing Dynasty.

from Zeng Guofan. As a matter of fact, the recruitment system where soldiers were owned by whoever recruited them, a system that enabled the Xiang Army to grow its numbers so quickly, is something that ought to alarm modern organizations: They should reject it in their approach to governance. A number of factors contributed to the fiasco of the Chinese army during the First Sino-Japanese War, but among them, the military organization and culture based on "owning soldiers" was the most critical. When the Beiyang Fleet fought bravely, the commander of the Nanyang Fleet chose to stand by and watch. The majority of court ministers secretly expected Li Hongzhang's[5] strength to be consumed in this foreign war. After their defeat in the First Sino-Japanese war, Liang Qichao commented that Li Hongzhang literally fought the entire country of Japan single-handedly.

History is a mirror. This "owning soldiers" philosophy helped the Xiang Army rapidly expand its numbers to subdue the Taiping Rebellion. Right about the same time, the disintegration of the Qing Dynasty loomed around the corner.

Soldiers are meant to be used by the general, but not to be owned. In its early stages of development, Huawei's management system bore superficial similarities to a conscription system, that is, "people are for you to recruit, and business is for you to do." But since the company was founded, Ren Zhengfei has been on high alert against factions that might lead to larger disunity within the organization, including things like absolute loyalty to superiors and associations between alumni or people from the same hometown. Huawei's frequent rotation of executives, the likelihood of them being assigned to higher or lower posts, and the "revolving door" system that sends them to work in different departments have to some extent reduced the likelihood of disruptive factions forming during the company's development. The collective resignation of the sales department, if you look back on it today, is no doubt a highly visionary move.

Then who "owns" the soldiers? The founding leader?

Of course not. Anyone in an organization belongs only to the organization itself. Ren Zhengfei doesn't make a single friend in Huawei, he doesn't

---

[5] Li Hongzhang (1823–1901)—A well-known politician in the late Qing dynasty, former commander of the Huai Army, founder and commander of the Beiyang Fleet.

privately grant public rights, and he doesn't engage in favoritism. Everything he does is customer-oriented and based on contribution. These actions present a clear answer to the question of ownership.

The third explanation behind the mass resignation is that Huawei was trying to go global. Ren Zhengfei said:

> The spirit behind the collective resignation of the sales team, I think, is more important than the action itself. It's a milestone on our path to internationalization. At the same time, it prompted a major adjustment of our management team, one that would prevent people from feeling lost, taking something that might have hindered us and turning it into something that drives us forward. Catalyzed by the actions of the sales department, all other departments will follow suit. If our managers and our organization lack a global perspective, then the company will never become a global company. It's normal for managers to move up or down the ladder. Just because you've dropped doesn't mean you can't climb up again, and just because you got promoted doesn't mean that you're going to rise up any higher. Only those with practical, on-the-ground experience can lead an office; only those with experience in a related field can lead a department. Managers must be deeply involved in what's happening on the ground in order to unearth new clues for development. Practice is the only test of truth.

Clearly, the collective resignation of the sales department was only the beginning. The decision makers wanted to shake up the whole company, to create a sense of urgency when facing the market and its customers, and more importantly, to gear all its managers towards global development. At that time, Huawei had already established contact with the IBM advisory team.

As noted, the IPD transformation was revolutionary in nature and it would reshape Huawei from the inside out. Where would they experience the greatest resistance? From those who had lost part or all of their power. The IPD reform challenged the traditional standard of power in the organization. In retrospect, the collective resignation along with the Huawei Charter (which was created almost in parallel) had been conducted as a prelude to broader institutional and procedural reform across the entire company. It was an organizational stress test, a rehearsal before the bugle call. From a staffing viewpoint, it was also part of a very real battle: trying to find the right lineup for the massive undertaking to come. In a sense, Ren Zhengfei had relieved some of the sting from the upcoming IPD transformation, so that when the time came, the American shoes wouldn't pinch quite so much.

One year later, the 70-member IBM advisory team arrived at Huawei.

Chinese-style Transformation: The Resignation of 7,000 Employees

In October 2007, approximately 11 years after the collective resignation of the sales department, Huawei once again staged a mass resignation event. This time, 7,000 people resigned in what the media dubbed "Resign-gate."

Huawei announced internally that all employees who had worked in the company for eight years or longer must go through voluntary resignation procedures before January 1, 2008, and then apply to the company for reemployment, signing a contract of 1–3 years.

According to Huawei's regulations, applications for reemployment could be made within six months after resignation, but the applicants would be subject to a review of their qualifications. Those who passed could stay, with the same compensation package as before. During those six months, the company would retain their shares—most of the employees back then participated in the employee shareholding scheme—after which the resignees would be allowed to cash out if they were not reemployed.

This was almost a carbon copy of the event that had taken place 11 years prior, but this time the responses were strikingly different. The previous event had made massive waves within the industry and caused some international competitors to feel a faint sense of looming pressure. They believed that a company with the guts to stage such a mass resignation and manage to succeed in spite of it must have a formidable team and a bold leader. "Huawei will outpace everyone else; it's only a matter of time," an executive at Motorola China once said.

But the 2007 event subjected Huawei to an avalanche of outside criticism. Hundreds of media outlets, both Chinese and foreign, covered the event with outraged headlines:

Massive Resignation: Huawei Challenges the New Labor Law

China Trade Unions Extremely Concerned about Huawei's Mass Resignation

Huawei Fires Its First Shot at the New Labor Law

The Internet was also alive with damning criticism.

Huawei was caught in a crisis and had managed to prove again that reform is usually a forced option. Beginning with the IPD reform in 1998, Ren Zhengfei had been emphasizing the need for an evolutionary approach to reform and gradual improvement aimed at optimizing management. The end goal was to get as close to a reasonable system as possible. So why did he launch two radical changes in human resources management, for which he was later called a "revolutionary entrepreneur"? He preferred conservatism to innovation in his business philosophy, so why did he make such a shocking move, which caused his fellow entrepreneurs to view him as a radical business leader?

The circumstances had changed. Starting in 2006, major mergers and acquisitions began to occur in the global telecom industry. Alcatel merged with Lucent Technologies, Nokia merged with Siemens, Sony merged with Ericsson, and Ericsson acquired Marconi. This rattled Huawei's nerves just as it was about to sit down for a nice, relaxing cup of coffee. After 20 years of rivalry between the mice and elephants, the elephants began to grow weary and sick one after another, while the mice had grown strong as lions. In 2007, the only companies that could really pose a threat to Huawei were Ericsson and Cisco. And at that time, the elephants made an unexpected (although predictable) move: They started banding together, warming each other up in preparation for an attack.

As a Huawei executive remarked, "For 20 years, we've had nightmares almost every day. If you choose to stay in the telecom industry, don't expect a good night's sleep."

Huawei's executive office issued an "Executive Summary on the Recent Human Resources Reform," which described the competitive landscape and explained the rationale behind the massive resignation of 7,000 employees:

> Both history and reality have shown us that, in the war of the global telecom market, like in any other war in times of peace, no company is able to win all the time. The balance can tip or suddenly reverse. Many world-class companies have been forced to cut jobs in order to survive, and some have disappeared into the storm. The road ahead is murky, full of uncertainties, and no company can ensure its long-term survival, nor promise its employees a lifelong career. And no company can tolerate lazy people either, because laziness is unfair to the dedicated workers and true contributors out there—tolerating laziness won't incentivize them, but quite the opposite. No great pie is going to drop from the sky and feed us all; you have to work hard to build hope for the future. There is no other option.

For many years, Ren Zhengfei worried about his team growing lazy. This was his greatest fear. He was the company's most vocal advocate for the spirit of dedication, and his greatest hope was that his company would remain lively and passionate.

What sets Huawei apart from its Western counterparts is its corporate culture. At Huawei, over 20 years of institutional reform have revolved around the core value that those who contribute more will be rewarded more. This is what has enabled it to rise when some of its competitors fall.

Although Huawei's culture kept it strong, there was another side to the company's situation: The new Labor Contract Law of the People's Republic of China had recently been passed. Under Article 14 of this new law, it was decreed that the employer, unless otherwise opposed by the employee, shall sign a labor contract without a fixed term with employees who either have worked for 10 consecutive years at the company or who have signed two fixed-term labor contracts with the company. Essentially, a permanent, unbreakable contract for employees who have worked at their companies for more than 10 years or who have signed two separate contracts of two or more years with the same employer.

This was a heavy blow to Huawei, which was in the middle of implementing its human resources reform. As part of its reform program, Huawei had adopted a new compensation system. Starting in 2006, its employees were compensated on the basis of their responsibilities and contributions, not on the basis of seniority, and every job was evaluated and ranked according to these responsibilities and contributions. People were paid on the basis of this job ranking and the amount they were given changed if they took on a different job.

The new compensation system would benefit dedicated, driven employees who actively shouldered responsibility and who made real contributions; while those who had become complacent or who took one long nap atop the laurels of their past success would be whipped into shape. Any senior employees who had lost their drive or ambition would be removed from their positions. Through this new compensation system, Huawei had hoped to preserve the dedication of its employees and motivate them to achieve even more as the company continued to expand its global presence.

The deep end of the reform, so to speak, was that it allowed two-way selection: employees could choose not to work for the company and the company could choose to get rid of employees who contributed less than they were paid. In essence, the company would not promise lifelong employment.

This reform obviously didn't comply with the upcoming Labor Contract Law. No one could tell exactly what or how Ren Zhengfei and the other senior executives felt about it, but it's safe to say that there was a heavy sense of anxiety.

On one hand, Huawei had entered the most competitive sphere of the global market and it was narrowing the gap between itself and its major competitors. Some Western companies were beginning to lose ground and were forced into an alliance to besiege Huawei, which was surrounded on all sides. Huawei didn't have any reinforcements, not to mention the national support structure that some of its Western competitors had. All it had to depend on was the dedication of its own employees. And it just so happened that the company's ability to sustain this culture of dedication depended heavily on a flexible human-resources policy.

The new Labor Contract Law was soon to be unleashed upon the world. If Huawei followed Article 14, its system of ongoing motivation would enter the ice ages. Huawei's leaders had dreamed of building a global company predicated entirely on free-market competition and this dream would soon be shattered.

They weren't deterred, however. The company moved forward with its planned reform and proceeded with the collective resignation. Huawei's lawyers had apparently done a good job. Although there was a surge of sensational hype, no laws or regulations were violated throughout the entire massive undertaking.

Under extreme pressure, Huawei had overcome yet another matter of life and death.

Why was it a matter of life and death? If its institutional reform had failed, Huawei's core value of dedication would have been lost and the company would have passed away on some foreign battlefield.

The Reform Here Is Nice and Quiet

The earth was quaking around Huawei, but the company had remained surprisingly quiet and calm. Not a single resigning employee had lodged a complaint; not a single dispute had been filed for arbitration or court ruling. Dozens of journalists who had managed to contact the employees in question were simultaneously disappointed and amazed. They had expected employees to vent their anger, feel bitter about the lack of job security, or fight to defend their jobs with the new Labor Contract Law in hand.

On the contrary, the resigning employees agreed with the rationale behind the reform: As it develops, a company accumulates a large number of senior employees who are well-paid but who have grown incompetent and have lost their passion. Without this reform, Huawei would have become a pool of stagnant water. With it, Huawei's people were shocked into a sense of urgency and the company could improve its market readiness.

Other employees who were not expected to resign were excited. "This reform put an end to the Employee Numbering System. We're really relieved to see the company scrap seniority. It makes us feel like our dedication actually means something."

Employee numbers were once a mysterious and holy symbol in the company. Ren Zhengfei's number was 01. Smaller numbers indicated greater seniority and quite often a much higher rank within the organization. But this system also divided employees, coloring the way they looked at and treated one another, causing new employees to suffer a feeling of inferiority.

After Resign-gate, Huawei rearranged the numbers of its employees and now Ren Zhengfei is around 110,000. Since then, every employee has been able to compete on a level playing field, thus creating a dramatic shift in the way things ran.

The change was surely dramatic: The mass resignation involved a total of 6,687 senior managers and employees, including Ren Zhengfei himself. The resignations of Ren and 6,686 others were approved by the board of directors in November 2007. Afterwards, the board retained Ren Zhengfei as the CEO of Huawei and reemployed 6,581 other employees. Another 38 employees chose to retire for age or health-related reasons; 52 found better career opportunities elsewhere; and 16 others were deemed to be unqualified and left the company after friendly negotiations.

It's important to note that some senior managers who were reemployed ended up changing jobs; some were promoted, some were down-graded, and some transferred to other functions. At the same time, a number of new and younger faces rose to the center stage in the years that followed.

Such a massive change affected the interests of everyone involved, but there wasn't any conflict throughout the entire process. External debate was at a roaring boil. Both the All-China Federation and the Guangdong Federation of Trade Unions sent investigators to Huawei, but the company was cool as a cucumber. People were reminded of two other events: the earlier mass resignation of the sales department in 1996 and the voluntary salary reduction of certain managers in 2003 during what Huawei calls "the Winter of IT." These were all radical changes, which some might label Chinese-style mass movements, and people were curious as to how they had proceeded so smoothly. How had Huawei's leadership succeeded in galvanizing the company with such a radical move? What was the secret?

Culture is the mother of all formal systems. Looking at the political systems of various countries, one might wonder why the United Kingdom and Japan have opted for constitutional monarchy and France for republicanism; most Middle East countries have retained autocracies while the United States has adopted a hybrid system of checks and balances that it repurposed from the best parts of various European systems. These differences are largely a result of different historical and cultural traditions.

Political systems are based on culture and each change to political systems occurs within culturally determined parameters. The romantic, radical, and dynamic French couldn't endure a monarch who lords it over everyone, passing down the crown from generation to generation, overlooking a motherland that's in constant pursuit of absolute liberty and fraternity. The British are sober, rational, and good at compromise; they naturally prefer harmony and balance between class distinctions and democracy, between the nobility and the commons, and between national symbols and actual politics.

Similarly, corporate systems are subject to something more fundamental. A state-owned enterprise belongs to the country and its managers are "gatekeepers" delegated by the nation. In theory, SOEs and state-owned assets in China belong to the people, including its employees; however, no employee can share dividends or exercise the power of an owner, such as electing or removing managers. Likewise, managers can't fire employees at will, let alone instigate mass layoffs in difficult times. This restriction has reduced the competitiveness of Chinese SOEs in the global market.

Conversely, there are often reports of corporate giants and financial institutions across the United States and Europe that fire thousands of employees at once. This is unimaginable for Chinese companies, including Huawei.

For over 20 years, Huawei has not fired employees on a massive scale. If the company were to one day slide into recession, would Ren Zhengfei or his successors follow the example of Western competitors and cut thousands of jobs? Would they be so bold? Could they bring themselves to do it? Ren Zhengfei and his successors certainly don't lack the courage to make big decisions when necessary, nor do they back away from hazards. But the problem is whether the company and its employees would follow the spirit of contract as those in Western companies do, given the special characteristics of Chinese culture that determine our collective integrity in the workplace. And would the Chinese government allow such massive job cuts?

Sooner or later, the day of reckoning will arrive. So in order to delay its arrival, Huawei has continued to transform itself and to embrace self-criticism in order to bolster and strengthen its corporate culture.

Inspiring dedication is the core tenet of Huawei's corporate culture and the employee shareholding scheme is one manifestation of this tenet. At Huawei, over 80,000 employees hold shares in the company. From a shareholder's perspective, why would anyone object to a policy that might help the company perform better and grow larger? Why should they care if they are downgraded in the corporate hierarchy, lose power, or are transferred to a lower-paying job if there are competent younger people who can create more value for them?

This was the general sentiment at Huawei and it largely explained why the radical, collective resignation of 7,000 employees didn't seem to put a damper on anyone's day.

To put it simply, 6,686 different bosses—both big ones and small ones alike—were led by Ren Zhengfei, the biggest boss of them all with a mere 1.42-percent share of the company, in a revolt against themselves. And the aim of this revolution was to stimulate the organization and mobilize each member against complacency and imminent crisis.

One can go one step further and conclude that the mass resignation was in fact an act of sacrifice among the 7,000 shareholders, done in order to protect their own vested interests. Each of them was an employee and, at the same time, the employer; there was no conflict between the two. In this sense, the media was barking up the wrong tree.

Art of Change: Timing, Tempo, Cost, and Other Factors

Ren Zhengfei: The Architect of Reform

All idealists are obsessed with the question: How is it that a fine and healthy dragon egg can crack open to produce a flea? Or if a baby dragon emerges, why is it marred and deformed? Similarly, many entrepreneurs are puzzled about why their companies got caught up in chain of misery or failed altogether, while other companies like Lenovo and Huawei have succeeded with management reform.

In the final analysis, while the outcome of reform depends on corporate culture, the methods and skills applied during the reform process are also critical. In the late Qing Dynasty, throughout the major reform movement of 1898 known as the Hundred Days' Reform, Emperor Guangxu had a severe lack of patience. In less than 100 days, he issued over 40 new orders and regulations and removed dozens of cabinet ministers, which shocked the entire court and the royal family. Regrettably, the reformers he depended on were a handful of lower-ranking officials who could write up a storm but didn't have any practical power. They had no sense of priority throughout the reform process and didn't unite everyone that they could have to support the cause. They weren't skilled at compromise, were incapable of striking a balance when needed, and lacked patience. They lacked determination too.

Influenced by these reformers who lived and died by the letter, the emperor became restless, undergoing act after act of sudden reform that flew in the face of the Empress Dowager Cixi, who might have otherwise been a strong supporter of change. The reform soon turned into a one-man show featuring the emperor and a handful of unsophisticated young men. He was isolated in court; even officials who supported or sympathized with his reform remained on the backbenches. The end of the reform movement was announced by shedding the blood of the reformers. This outcome hadn't come about by chance and conservative power was only partly to blame.

About a century later, Deng Xiaoping championed another reform movement and he was wise to past mistakes. Although his reform was carried out in response to an entire nation that was ready and waiting for change, Deng's great mastery of the art of reform can still be looked to as a highly valuable source of inspiration for all leaders, business and political alike. He liberated and reemployed a large number of ousted officials, turning them into the leading force behind the reform movement. In parallel, he started a movement to liberate people's minds and create a favorable ideological environment for the reform. He also nurtured a class of intellectuals who justified the reform and called for nationwide support. To guarantee smooth implementation, Deng reformed the armed forces. He then drove a series of pilot projects in the countryside to stabilize the country at the lowest, yet most fundamental level of the economy: the household-based land-contract system for the farmers. After everything was up and running, he insisted that the process be gradual. He called the country to feel around for stones as they crossed the river, or to put one foot on the accelerator and the other on the brake, so that the vehicle of the country could traverse ridges, avoid holes, and get around corners. By 1997, when Deng Xiaoping passed away, China was a mere 20 years into its reform, but it had already begun to see different levels of achievement across all sectors.

A large number of people in China's business community call themselves pupils of Deng Xiaoping, including Liu Chuanzhi, Zhang Ruimin, Wang Shi, and the list certainly goes on. They all share certain traits: resilience, patience, and vigilance. As the boss of a private company put it, "We are like mice. We hide ourselves in the hole when the wind is strong outside, and come out after we know the wind is gone." Moreover, they are tolerant and inclusive. They pull people together and accept everyone in spite of their defects or problems. At the same time, they stand by their principles, almost ruthlessly. They compromise, but they don't forsake their core beliefs. They are decisive, self-aware, and they hold on to a stubborn faith: As long as they are businessmen, they will never dabble in politics.

Interestingly, these entrepreneurs all started their businesses around the age of 40 and were around 50 when their companies began engaging in major transformation. For the leader of a sizeable organization, this is considered a mature age. At least this seems to be the case in China.

Ren Zhengfei's role throughout Huawei's 18-year history of reform has been more like that of an architect. From the Huawei Charter to IPD and the other five structural reforms, from the mass resignation of the sales department to the 7,000-person Resign-gate, you will find that, here and there, he reveals information about every upcoming reform in the speeches he gives, usually around two years before each change occurs. His perspective becomes clearer and clearer as each reform is about to enter the launchpad and when the change officially takes off, management at all levels is just about mentally prepared for it.

But Ren Zhengfei's ideas are always in the form of metaphysical speculation or matters of principle, rarely involving specific suggestions or recommendations about how reform should be carried out. He might not be a professional "top-level designer," but his intuition and shrewd insight into organizational and interpersonal relations is extraordinary. Perhaps the contents of the Huawei Charter have long been forgotten by Ren Zhengfei and his management team, but in the course of 18 years, any Chinese or Western-style reform (mostly a mix of both) has invariably fallen into the general framework of the charter—a rare phenomenon in the evolutionary history of Chinese companies.

In addition, at different stages of organizational development, Ren Zhengfei has been able to perceive the symptoms of problems long before they break out on a large scale. In 2009, he spoke out loud and clear: "Let those who can hear the gunfire call for artillery." In 2013, he further specified that the company needs "simplified management." And now, carrying out reform to address the management problems of large organizations has become a long-term mission for Huawei as it looks into the future.

Set the Chickens Flying and the Dogs Jumping

Reform is aimed at stimulating the vitality of an organization. For companies, the biggest challenge doesn't come from the outside; changes to market conditions and the strength of their competitors are not the greatest threats. What companies suffer from most often is internal stagnation or rigidity as a result of organizational friction, laziness, and fatigue. Therefore, reform is necessary to stimulate the organization and delay its descent into degeneracy.

Successful reforms are based on the insight and prudence of those who lead them. Political or commercial reform can't be accomplished by one person or even a handful of people; it requires powerful leadership and a resourceful group of advisors. Reform led by a firm-handed leader with the help of a group of dedicated professionals is much more likely to succeed.

And reform is a systematic project. Any change sets off a chain of events. The goal of reform is to ultimately clean up (not in the political sense) or transform the organization. Prioritizing targets is crucial, however; it's a bad idea to try and tackle every problem at once. Any sudden moves tend to spook the farm—set all the chickens flying and the dogs jumping about. In other words, sudden change across too many dimensions at once can send the organization into complete disarray, causing enemies to rise up in all directions. Another pitfall is when organizations assign their resources evenly to all reform targets, not considering the fact that their resources are limited. When the scope of reform is too broad, all efforts typically come to a grisly or abortive end. The Hundred Days' Reform is a grim example of this.

Of course, the whole point of reform is also to get the chickens flying and the dogs jumping in order to reinvigorate the organization. But when this happens, the reformer has to observe the implications and direction of unrest. By nature, reform inevitably hurts some people, especially those who are deprived of vested interests—those who naturally hop about in fury or violent objection to change. But if this type of resentment becomes too widespread, or worse, if the chickens and dogs join forces to oppose reform efforts, the risk of failure will increase. On the other hand, reform is more likely to succeed when it is carefully planned and sequenced—starting with experiments to drive tangible results and then riding on the momentum of those results to roll out the program completely.

Huawei's six reform projects began with IPD, followed by supply-chain and human-resources transformation. Then in 2007, the company's finance system became the next object of reform. At a meeting to present Huawei's plan for financial transformation, Ren Zhengfei said:

> I have never supported financial transformation that begins with the budget system. If our operational system is still a mess, how can we budget correctly? For now our budget system has to move ahead on its own; we must plow the soil and level the ground before our advisors bring us the seeds.

Even the IPD reform program was not launched throughout the entire R&D organization to start with; rather, Huawei implemented it in a new project group of wireless products that involved fewer than 100 people. The benefit of experimenting on a small scale was that backlash could be minimized if the program failed; if it succeeded, all the participants in this program would be encouraged by the results at that "phase" and a broad demonstration effect would follow. This was the first lesson that Huawei learned from experimental reform.

The second lesson was training. After the reform plan was mapped out, large-scale comprehensive training sessions were held for executives from different departments, including R&D, sales, supply chain, finance and others. The training program lasted for four months, training a total of 3,700 managers who were at level-three positions or above, with each training session comprised of 200 people. These systematic training courses were organized by the reform steering committee on weekends. After each session, participants were required to write down their takeaways, which would then be commented on and published in the company's *Management Optimization* newspaper. Internally, this type of "large-scale brainwashing" that's done in preparation for change across the entire organization is called "cleansing the salt from the soil of the brain."

The third lesson was not to begin transformation projects in business departments that are critical to the business and that appear to be successful. For example, C&C08 digital switches, the company's knock-out product at the time, were developed by the fixed network product line, so IPD transformation began with wireless products. The program met with the least possible resistance in the wireless product line and in the Beijing and Shanghai research centers, and once it was fully on its way, the effectiveness of the program was far more visible. This also explains why a large portion of the R&D management team and even the corporate management team have come from the wireless department.

Reform Comes at a Price

The massive resignation of 7,000 Huawei employees was labeled as an earth-shattering act of reform by the media, but the world was more shocked by how quiet the company had remained. This was due, in part, to the fact that the 7,000 resignees were shareholders; it's also because Huawei paid a hefty price for this seemingly radical change.

For employees who have worked with the company for over eight years and who have resigned and then applied for reemployment, Huawei adopts an "n+1" compensation policy, where "n" is the number of years the employee has worked with the company. For instance, suppose an employee makes CNY12,000 a month and earns a year-end bonus of CNY120,000 (a monthly equivalent of CNY10,000); if he or she has worked at Huawei 10 years, the compensation would be CNY22,000 (monthly salary + monthly equivalent of year-end bonus) times "10+1," or CNY242,000.

Over the years, this has cost Huawei over CNY10 billion. Of course, Huawei has carefully weighed the product of this investment. What's more important and valuable than the sustained vitality of the organization?

Compensation for resigning employees aside, over the past 18 years, Huawei has spent more than CNY30 billion on its six major reforms, including IPD, ISC, and IFS. As one of IBM's largest global clients for consulting services, Huawei has paid more than CNY1 billion in consulting fees to IBM alone. According to a former mid-level manager who participated in running the IPD program, it seems that Huawei never set a limit as to how much could be spent on a given transformation: money always goes where there's a business need. In other words, the purpose is always more important than the expense.

This has certainly contributed a lot to Huawei's continued growth over the years. When the ISC reform was launched around 2002, the company's business performance declined, and in response, the ISC program was simplified and downsized. As a result, the program was not as effective as the company had expected. This, of course, just offers one perspective; there are divergent views even today as to why ISC didn't fully work out. Some, for example, argue that the experience and capabilities of the consultants were also a factor.

For 18 years, Huawei's annual budget for transformation projects has remained at 1–4 percent of its sales revenue, averaging around 2 percent.

Consulting Is Not the Same as Reform

Why is Huawei so willing to endorse IBM? The reason is very simple: because IBM showed Huawei the ropes, helping it to hop on the globalization train with a set of Western systems and processes. But as an IBM consultant pointed out, consulting is not the same as reform. IBM was neither the reformer nor the target of reform. Huawei was. The success or failure of reform fundamentally depends on the boss of the company, on his will and determination to push through against all odds.

Ren Zhengfei's persistence has played an important role throughout Huawei's process of reform. According to Arleta Chen, Ren was convinced that Huawei had to learn from the United States if the company was to globalize. In addition, between being technology-driven and customer-driven, he chose and adhered to the latter as the company's guiding value. He has never wavered on these two principles, nor would he ever allow anyone else to go their own way.

As per Arleta Chen, "apart from this, he's basically hands-off." Arleta Chen and other IBM consultants rarely saw Ren Zhengfei participate in the discussion of reform plans; in fact, they rarely saw him at all. He never intervened in the planning and implementation of specifics. A Huawei executive commented on Ren's approach, saying, "He always stands on the shore, watching how you swim. Sometimes he doesn't watch at all, just points out the direction and waits for you to deliver the results. Essentially, he's quite detached."

Interestingly enough, though, he pays close attention to certain details. Li Aixin once served as director of the IPD program and later joined IBM to work on their management consulting team. Li recalled,

Every time the boss saw me, he would ask if I had paid the consulting fees. Sometimes when payment processing took a bit longer than usual, he would get pissed off. Mr. Ren was not doing business with consulting firms; he was hiring a teacher for Huawei.

When someone complained about paying too much, Ren Zhengfei replied, "Don't be foolish. You are paying US$680 an hour, but you're getting the knowledge they've developed over 30 years. If you ask for a discount, they'll only hand over knowledge from the past three months. Which one is a better deal?"

In order to make Huawei's IPD program gain more attention and support from IBM headquarters, Ren Zhengfei decided to replace all of the servers and application software that Huawei purchases with IBM products, and he kept issuing these instructions at every meeting he attended, even though "IBM's equipment at the time was expensive and not that great."

Huawei's process of selecting consultants is very strict. First, a list of recommended consultants is provided by the consulting firm for Huawei to choose from. One important criterion is that all candidates must have a successful track record in real business. Fei Min, a retired member of the executive management team, was once responsible for the IPD program and his viewpoint on the matter demonstrates the sober awareness of Huawei's leadership: "A consultant without business experience is at most a counselor.[6] Their knowledge is learned but not experienced."

Huawei's people are thirsty for knowledge and they are strong learners, which are both important factors in ensuring the smooth implementation of reform. In the face of change, their typical reaction is as follows: For things I don't know, you teach me; but at the end of the day, the reform program is my business and I have to pull through no matter how hard or exhausting it is. As per Li Aixin, "change will certainly make the organization and individuals uncomfortable, but there's no turning back."

---

[6] Translator's note: The word he uses here is *shīyě*, which refers to a special type of advisor in ancient China. There is no direct equivalent in English, but *shīyě* played the role of viziers, clerks, and advisors, adept at wordsmithing, secretarial duties, and theory, but with little practical experience in governance and war.

This is the tie that binds Huawei's people with their consultants. When the consultants walked into Huawei, they didn't have a pocket book filled with readymade solutions. However, motivated by the curiosity of Huawei's staff, they came up with a living solution that fit perfectly with Huawei's business. It's like military drills on the drawing board. The drawing board came from IBM, but the soldiers were from Huawei. According to Li Aixin, "for consulting companies, consulting only gets about 10 percent of the job done, whereas reform itself takes more than 10 times the effort."

Sun Yafang concluded that the key to reform is to be the vehicles of change; consultants only provide a framework and their experience. In a successful reform, solution design only accounts for 30–40 percent while implementation carries a weight of 60–70 percent. Most companies pay consultants for the design, but Huawei mainly spent on the latter. According to Guo Ping, implementation is the real beginning of a reform program. He recalled, "In the IPD program, the consultants even taught us how to build a template, how to draw a table and fill it out, how to make a meeting more effective… [E]verything was hands-on."

Through the IPD program, Huawei's quick-learning staff mastered a whole set of reform methodology, including why to change, how to change, how to make change sustainable, and how to implement it. Through the program Huawei also developed its own mature team of reform experts. By the time the IFS program kicked off in 2006, although IBM consultants were still brought on board to help, the whole process of design, promotion, and implementation was led by Huawei itself. This time IBM was only being consulted as an advisor.

Meet the Buddha, Worship Him: Meet the Devil, Kill Him

Is there a strong resistance to change in Huawei? Some believe that resistance starts out as a problem with the head and, over time, it becomes a problem of the butt. Essentially, where you sit will determine what you think.

One time, a meeting with Ren Zhengfei left a very strong impression on Li Aixin. In the meeting, Ren took out a utility knife and said to Sun

Yafang, the company's chairwoman, "If you think the IBM shoes pinch your feet, then cut your feet off." In fact, Sun Yafang was the staunchest advocate of IPD transformation and also the director in charge of this program. "If her feet can be cut off," Li Aixin thought, "I doubt we'll be able to keep our heads if we object to change." From this experience, Li got a feel for how strongly Ren was truly bent on reform.

At a meeting in 2013, when discussing resistance to the ongoing IT transformation, Ren Zhengfei said to Deng Biao, president of IT department, "Let me give you a knife—either kill those who stand in the way, or kill yourself." It's clear that change in Huawei, like in any other organization, is by no means easy, but the determination and the willpower of Huawei's leadership is also starkly clear. At the same time, this determination speaks to a cruel reality throughout history: In order to change, someone has to make a sacrifice.

So how does one properly brandish the knife? As a former Huawei executive (who asked to remain anonymous) described, there are two ways of using it. One is to "kill both the Buddhas and the Devils that stand in the way." The ones who adopt this approach are the go-getters who cut a bloody new trail ahead of them during reform, but who are often the first to be sacrificed after the program has succeeded. "The company should treat these people nicely."

The other type of person does it differently. "When they meet the Buddha, they worship him; and when they meet the Devil, they kill him." Even with a sharp and formidable knife at hand, they are good at distinguishing between different groups of people who might stand in the way. If the problem is a matter of personal conception or point of view, these people will go all out to explain and communicate the problem away. If the resistance is associated with vested interests, they strike down like a thunderbolt. They know when to be flexible and when to use the iron fist. "The boss likes both types, but the second type—those who have both principles and strategy—are the ones with the potential to become future leaders."

Overall, most of the managers, experts and employees who have participated in Huawei's reform programs have been rewarded in proportion to their contributions. Some of them became company executives while others have gone on to engage in change management for years. In the Whiz Kids award ceremony in 2014, quite a number of employees won the top prizes for management. But many of those who directed and implemented the IPD and ISC programs had already left the company, because more often than not, when the reforms wrapped up, their positions in the original department were lost to someone else.

Reformers at Huawei are the unsung heroes. They take on no less—if not more—pressure, grievance, frustration, and failure than their fellow colleagues in sales or R&D, yet their contributions are a lot more difficult to measure. Before the progress review meeting of an IFS program milestone began, Meng Wanzhou, Huawei's CFO, hugged each and every one of the 200 team members, including the IBM consultants. Most people shed tears. It was quite a revealing moment, shedding light on the difficulty and pain behind the scenes.

Positive affirmation and recognition of the people who make reform happen is a key factor in ensuring the ongoing success of any reform.

Self-criticism Goes before Reform

West Point superintendent General David Huntoon pointed out in a recent speech that, in the 21st century, critical thinking is essential for successful military leaders. Ren Zhengfei agreed and commented, "That fits very well with our management philosophy. To evaluate the prospects of Huawei through critical thinking, innovative thinking, and a historical viewpoint is a heavy task that Huawei's new generation must shoulder."

All reform comes at great price, both explicit and implicit. An incautious approach to reform can also lead to a long-term price. By contrast, self-criticism is the least intrusive tool—it can be used repeatedly and it comes with the lowest costs. To a large extent, Huawei's success is the result of an interwoven application of self-criticism, innovation, and reform.

Criticism is a gradual and progressive tool that keeps an organization healthy, just like running water that doesn't go stale. Regular, purposeful, and systematic organizational self-criticism has become a tradition at

Huawei. But when criticism isn't enough to eliminate the corruption and laziness of the organization, reform needs to take the stage and start doing what it does best.

Criticism goes before reform as a warning against organizational disease.

In fact, reform can't exist independently of self-criticism. The two function side by side, and sometimes criticism can help clear the way for reform. Both are essential for any company. Just imagine what might become of a company where everyone seems to get along all fine and dandy, never exchanging a single unkind word, when suddenly it experiences the shock of massive, disruptive reform?

Moreover, an organization that is unable to embrace self-criticism or isn't resolute enough to do so is unlikely to succeed in institutional reform when things get real.

Criticizing the Whiz Kids

The Cold-blooded Surgeons

Half an hour after the first award ceremony for the Whiz Kids ended, Ren Zhengfei had an interview with around 20 domestic media outlets. The first thing he said was, "I'm here today to criticize the Whiz Kids." Unfortunately, his opening remarks didn't draw much attention from the audience.

Undoubtedly, Ren Zhengfei is a calm and profound man, one who doesn't let one inclination cover up another. While admiring the fact- and number-based rationalism of the Whiz Kids, we shouldn't neglect the purpose of management.

In my lectures to Huawei's staff, I suggest that *The Whiz Kids: The Founding Fathers of American Business—and the Legacy They Left Us* (Chinese edition), a 680-page classical textbook for management, should be read as two separate parts for study: The first 300-plus pages illustrate the glories and victories of management based on numbers and reason, while the second 300 pages criticize the blind spots and deficiencies of this approach.

The author speaks highly of the Whiz Kids' achievements: they brought Peter Drucker's ideas of management innovation into the mainstream business world. Through legend and hyperbole, they also established themselves as the prototypes of professional managers in the minds of millions. The author praises how they were able to organize and run gigantic companies, build business empires, and manipulate the government: essentially, how they were able to create a perfect world based on logic, reason, and rationality. On the other hand, they had a blind and firm belief in the notion that something that had worked before would keep on working forever. They had attempted to play god, but in doing so, they neglected the responsibilities of regular people. The author also quotes a remark from a well-known financial media outlet about how systematic analysis was like a religion to the Whiz Kids, who believed that statistics are a victory of quantitative methodology, that quantitative methodology was a victory over the loss of life and over death itself.

The Whiz Kids were a group of outstanding, professional management experts, but they were not army generals or commanders born in the heat of battle. They lacked a sense of tempo, so to speak: While they had saved Ford from chaos and brought it to a state of prosperity by means of numbers and procedures, they confined themselves to their doctrine and philosophy of perfectionism, pushing Ford to such extremes that their over-management nearly choked the company. They controlled costs to an excessive degree, which killed innovation, and their restrictive corporate culture led the company into a recession.

Adhering to a philosophy of ruling by numbers in the office, the Whiz Kids tried to use sandboxes and diagrams to completely cover up the relevance of real-world experience in the corporate battlefield, which enabled them to release successful products based on market research data. At the same time, however, to realize the blueprint in their minds, Jack Reith, the "super adventurer" among them, went so far as to forge data to launch a product which didn't reflect customers' needs at all and led to huge losses.

Another Whiz Kid, Robert McNamara, when he became the Secretary of Defense during the Vietnam War, told journalists that, in spite of all the defeat the United States had experienced, all the data indicated that the United States was sure to win. McNamara was voted as the greatest contemporary US entrepreneur in 1967. As the media noted, his proficiency with numbers and addiction to facts were the causes of both his success and his failure.

Tex Thornton, the head of the Whiz Kids, founded Litton Industries after leaving Ford. A forerunner of mergers and acquisitions in the United States, Thornton completed 25 business acquisitions within eight years. However, because he tended to pursue growth for the sake of growth, the company became subject to the whim of the capital market, driven by ambition far stronger than Earth's gravity. Litton Industries ended in a tragic debacle.

At Ford, the Whiz Kids established a powerful financial control system, with as many as 14,000 staff at its peak. The auto giant had 300,000 employees and this system provided a reliable surveillance platform for cost control and performance management. But because the system went to the extreme, supervision became an end of its own and, possessed as it was with paramount power, the finance department became the de facto center of control at Ford. As a result, the organization began to grow distant from its consumers, its management system grew rigid, and the company began to lose its competitiveness.

The deep flaw of the Whiz Kids was that, although they were critics, they never criticized themselves. Brilliant successes continued to roll in, piling up into a mental blind spot. They began to believe that tools could make them invincible. As a result, they got puffed up with ambition, grew fearless, and became a group of delusional workaholics. Almost every one of them walked right into a closed-up fortress of their own making and, eventually, most of their careers met with tragic end.

After the last Whiz Kid retired from the company, Ford did two things almost immediately: The company cut its finance team in half, from 14,000 to 7,000, and then arranged for its managers to attend a training course—Irrational Management.

A few years later, Ford had resurrected.

Just around the time that the Whiz Kids retired, the American management science community was reflecting on one question: how to quantify morale, discipline, leadership, innovation, integrity, and courage in an organization?

Management can't be fully captured with such linear thinking. In any organization, there must always be at least two opposing forces and opposing ideas.

Now back to Huawei. According to Hu Houkun, during Huawei's HR reform, he had reported to Ren Zhengfei several times, but Ren never let on to what he was thinking. It wasn't until Hu told him, "Boss, we're still a Chinese organization, but we're using a Western approach," that Ren seemed to be relieved: "Good, that puts my mind to rest."

The Purpose of Reform Is to Get More Food

When Ren Zhengfei began to stray away from the script during his speech at the Whiz Kids award ceremony, the first thing he said was: "The purpose of our reform is to get more food."

Like the question of the chicken and the egg, the relationship between process and purpose has always been a paradox without a definite answer. However, organizations must hypothesize about this relationship at each phase of their development. As Ren Zhengfei once said, "Hypothesis is man's greatest mode of thought."

Here is Huawei's hypothesis: Purpose is above all else. The sole criterion for measuring the success of a reform is whether or not it promotes productivity. The IFS reform went on for eight years and it aligned perfectly with this principle.

Huawei began its financial reform as early as 2004, but the focus was only on the accuracy and timeliness of accounts receivable and payable. Back then, Ren Zhengfei had already proposed that finance needed to integrate more closely with the business side of things; otherwise, it simply couldn't add value. In 2009, at a workshop on IFS reform, he said for the first time that Huawei needed to deal with the uncertainty of results by leveraging the certainty of rules.

This became the guiding principle for the company's IFS reform.

To a great extent, the reform of the finance system was a type of *business transformation*, which would inevitably lead to change and adjustment in business operations and in the way that business leaders behave and think. As the reform progressed, fibers of the finance system would extend to every corner of the business, like a cobweb. Would employees in the field be ready to accept it? Or would they fight against it? Would the reform affect the company's expansion? These were among the questions that were raised and constantly magnified during the design phase; as a result, the IFS reform proceeded rather slowly in its early days. However, there was indeed something rational in their skepticism.

The IPD reform was a program that focused predominantly on R&D. Indeed, R&D is the heart of an organization and the IPD reform was designed to enable that heart to fully integrate with the brain, the skeleton, the nerves, and the endocrine system, all around the nucleus of the organization: its customers. In the final analysis, however, IPD was an internal reform within the organization and its design scenarios were not that far-reaching and complicated. It was therefore a lot easier to implement than the IFS and LTC (lead to cash) reforms. Both the IFS and LTC programs touched on the global market: a market that, for Huawei, was worth tens of billions of dollars and involved tens of thousands of people in 170 countries and regions, as well as many product lines. The strategic point of entry for the IFS team was figuring out how to inherit and reinforce success factors, how to minimize the existence of problems, how to mitigate the risk of failure, and how to balance business expansion with effective control.

Some argue that a visiting monk delivers better sermons.[7] This time, the "visiting monks" were still IBM consultants, but the design of the "scripture" (or the overall reform proposal) was led by Huawei, based on its own knowledge and successful practices while drawing upon the experience of IBM, Ericsson, and other industry players. Li Hua, one

---

[7] Translator's note: This is a well-known saying in Chinese, which means that the opinions of someone outside of a given group (a visiting monk) are taken more seriously than the opinions of those inside the group (in this case, the monks inside the temple). Some might argue that it's because they can see the truth from the outside; some might argue that it's because people are more likely to listen to strangers as opposed to those closest to them.

of the program leaders, put it very vividly that IFS was like "making a Chinese version of Marxism—it's something with Chinese characteristics. We put that Western experience through the old Huawei crucible, and forged us something unique."

The gist of the reform's "scripture," complete with Huawei characteristics, was captured in a few key messages: Control is for the sake of better business expansion; the objective is to enable the profitable and sustainable growth of the company through accelerated cash inflow, accurate revenue recognition, visibility into project profit and loss, and well-controlled business risks. The reform would help the finance department transition from a purely functional organization to a business partner and value integrator.

The transformation would begin at the front lines, where they had the most pressing business needs. Before the reform, invoicing accuracy was only 70 percent and there were serious discrepancies between the accounting book and actual fulfillment. After the reform, invoicing accuracy rose to above 95 percent. The representative offices where the IFS pilot programs were undertaken commented that, before, when they signed a contract, they would always worry about not being able to collect on accounts receivable. The reform helped them resolve a massive headache.

As the reform had been designed with the needs and pain points of the frontline in mind, it won extensive support throughout its entire implementation process. The reform team, diverse as it was, had members from sales, delivery, R&D, finance, and other functions. The reform team was very willing to engage the frontline, knowing the pains, difficulties, and needs they experienced out in the field.

It turned out that Huawei's leadership mobility mechanism, or its revolving door system, had not only nurtured well-rounded managers but also offered a solid organizational guarantee for the success of its reform.

Top-level design is the first step of any reform and IBM is undoubtedly the master of architecting systems and processes. However, some people also argued that IBM put too much emphasis on top-level design and theoretical frameworks. The wisdom of experts was overvalued and "frontline staff were only there to execute." Such an approach was feasible and effective for Huawei in its developmental stages and throughout the transformations of its R&D and supply chain, but it was not entirely appropriate in 2006, when Huawei was already on its way to becoming a global company.

At Huawei, there was certainly a blueprint for reform at the top level, but the blueprint had to be grounded, adapted to what the frontline had learned as they navigated the reality of business, or as they felt around for stones while crossing the river. They had to identify where the blueprint converged with business on the ground so that the design would closely reflect market truths. That's why Huaweis frontline commanders, or business leaders, had gotten involved in the IFS program either directly or indirectly from the very beginning of the reform.

Not long after IFS was launched, Ren Zhengfei wrote to IBM's CEO, inviting them to be Huawei's consultants in this next round of reform. In a conversation with the CFO of IBM, Ren expressed gratitude to IBM for turning Huawei from a small company into a large one through the IPD reform, and hoped that through the IFS reform, Huawei would be forged into long-lasting company.

This time, though, there was a distinct difference. For the IPD reform, Huawei was committed to "wearing a pair of American shoes" and would "cut its feet" if the shoes were too tight. During the course of the IFS reform, however, there was no talk of cutting feet; with the change program, the company had already changed.

Beliefs, Systems, Leaders

Why do some well-managed, large companies collapse or wither away when nothing appears to be wrong? The management-science community has been discussing this question non-stop for the past few years, when in fact the question itself is moot. What are the criteria for good management? Can it still be called "good management" if the company has collapsed or if it's got one foot in the grave?

Beliefs are the source of systems. When they grow to a certain size, companies must have a set of systems and processes that really work.

This is the underlying logic of business management. However, more important than systems and processes are the beliefs—the company's values. Some century-old business giants in the West once boasted a rich legacy of systems and processes, so why did they begin to rot and malfunction? The reasons are many but, at a very fundamental level, it's because they didn't get their values right. They were all successful when they embraced the universal philosophy of "customers first," but whenever they turned their backs on this value—this iron law—they began to drop like flies.

Huawei has exerted a great amount of effort in building out its systems and processes, and yet they are still not adequate in many ways. This being the case, how has it managed to stay on course for 27 years? The key lies in the firm and sober awareness that lies deep in the minds of Huawei's leadership: The customer is the center of everything in the organization. Innovation, reform, and the development of systems have to be carried out around customer needs, both explicit and implicit. There's but one purpose: "to get more food," or to deliver real, tangible, and meaningful business results.

Strictly speaking, Huawei has not invented any concepts or beliefs uniquely its own. But when it comes time to translate those beliefs into systems and processes and eventually into every single step of corporate management, innovation and reform are indispensable. At many companies with the right corporate values—where they advocate customer centricity—they often fail to put the right systems and processes in place to deliver on the value proposition. This is for several reasons.

First, effective systems are only those that have been proven in practice. Each of Huawei's institutional reforms has been a combination of top-level design and their feeling-around-for-stones approach. Reform starts with the real world and ends in the real world. That's the way to make corporate systems into living things—something that's true to life and can really work and which can be widely supported by entry-level organizations and employees.

Second, systems and values have to connect and reinforce each other. If the systems and processes of a company emphasize control too much and not a balance between governance and motivation—particularly if they're not designed to promote customer success—then there will be a disconnect between the systems and the values. This disconnect may well result in confusion about the organization's identity and misapplied business management, and eventually lead to confusion among the company's managers and employees.

Third, strong leadership is necessary. Any change of systems or processes involves the redistribution and adjustment of power and personal interests. At every step, reform leaders must remain determined, have a strong will, employ simple yet effective wisdom, and act with simple and effective maneuvers, otherwise the reform will fall short—or die an early death.

Last but not least is innovative thinking. The world is rapidly changing: The farther a company goes, the more it needs good ideas, insightful thinkers, and strategists. No one has a crystal ball to predict the future, but when a group of thinkers comes together to develop a set of open-ended hypotheses about the future, the company will have a far greater chance of getting things right.

If we compare Huawei's development to a ballgame, it's clear that Ren Zhengfei's thought leadership has played a predominant role in the first 4–5 innings. For the next few innings, though, while Ren's wisdom is still needed, a team of exceptional senior leaders in the company will have to step up to the plate. They need to come up with new ideas to ensure that Huawei doesn't stray too far from its course or get lost amid the ongoing evolution of its corporate systems.

The future is cloudy. Uncertainties loom up ahead—all around us, in fact—and they might at any time draw an otherwise great organization into their murky depths. No individual or organization can afford to harbor the dangerous illusion of invincibility. In times like these, the power of thought and exceptional leadership are more important than ever before. The belief that a perfect system can replace people is simply unrealistic.

From this perspective, Huawei is no different from IBM, Ericsson, or Google; all of them face the exact same challenges.

Leadership at Huawei: The Essence and Application of Power and Influence

Chapter 9

The Human Nature of Leadership

A common belief exists that the most important talent issue of any organization is its leadership. The need for leadership is especially prevalent nowadays as business is conducted at a global level, interconnectedness is high, and rapid changes in technologies and industries are taking place. Under such circumstances, organizations need strong and wise leadership that acts as the catalyst that brings all other organizational elements together. Leadership that inspires, motivates, and facilitates the translation of the organization's potential and talent into business successes. To put it boldly, an organization without a leader can be compared to an army without generals—in other words, some-thing unthinkable. The army is indeed an organization par excellence that considers putting much effort into training their officers and develop-ing the necessary leadership skills to deal with any kind of change as a prerequisite to success. Huawei has been founded by a leader who without a doubt will: (a) agree with the focus an army has on leadership and (b) understand why it is so relevant to business nowadays.

Although Huawei is recognized as an employee-owned company, it is very clear that the DNA of the company is colored strongly by its founder Ren Zhengfei, who at all times and in different roles remains in control of the company. The achievements of Huawei so far have been remarkable and the journey to those successes has been one of ups and downs. It is only with effective leadership that this combination of successes and failures could have led to the status the company is now enjoying. So, what makes Huawei's leadership so exceptional? The aim of this chapter is to examine in a structured manner whether there are any secrets that we can uncover that make it easier to understand why Huawei has become a global leader since its inception in 1987. In doing this, we will make use of examples and insights discussed in depth earlier in this book to illustrate leadership specifics of Huawei.

The first important point to make is that, taking into account that only in April 1988 at the National People's Congress did China's parliament approve that private companies were allowed to do business, it is no exaggeration to say that Ren Zhengfei can be regarded as a prime example of a Chinese business leader representing the first wave of truly entrepreneurial business leaders. Together with many others, he had to face the mighty state-owned enterprises at a time where being a private business owner was looked upon with suspicion. Despite his struggle with the SOEs and the obstacles thrown at him by the protective government, Huawei always survived and with each step in its development, became a landmark in its industry. The leadership of Ren Zhengfei across those years has shaped at least in part the way people think about doing business in China. In fact, one could even say that the influence of thought leaders like Ren Zhengfei has led to a new business landscape in China where the government now expects innovation to come from the private sector and not the SOEs. For example, due to its focus on total service, going beyond the product delivery—compared to the narrow focus of foreign companies on the product only—combined with ever-increasing investments in R&D, Huawei has become the example for Chinese companies that proves they also can become innovative players on the world stage. What kind of leadership underlies such a significant transformation in the Chinese market?

In a 2011 *Harvard Business Review* article, Ikujiro Nonaka and Hirotaka Takeuchi argued that in a world with an ever-increasing pace of change, business leaders struggle to reinvent their corporations rapidly enough to cope with new technologies. They also added that as a result, those leaders are unable to develop truly global organizations that operate effortlessly across borders. According to these authors, the reason for this is that we have no wise leaders anymore. If we use their perspective on what leadership should achieve, we can say that Huawei has met almost all of these targets: dealing with changes in the industry and political environment and within this turbulence, emerging as a key global player. If this is the case, is it fair to conclude that Ren Zhengfei has fueled the development of Huawei because of his wise leadership? Looking at Huawei as an organizational structure, it is very clear that fostering wisdom and developing intellectual skills via different initiatives is an essential aspect of its existence. One initiative concerns the inclusion within the company of the Huawei University, where training sessions are held on how to conduct business at Huawei and provide solutions with a focus on serving the needs of its customers. A second initiative is the use of a value-training camp where new employees are introduced to the values that form the backbone of the growth and success of Huawei. The aim of the camp is repetition of the core values and instilling the lesson that both individual and collective growth is possible by serving the needs of the customers.

Looking at the more individual level, Huawei's focus on wisdom as a key aspect of continuous growth is without a doubt defined by its founder's love of studying and developing knowledge across borders. As John F. Kennedy once noted, "Leadership and learning are indispensable to each other," and Ren Zhengfei embodies this wisdom in many of the things he does. An important facet of this knowledge process is that Ren Zhengfei is known as a storyteller who regularly talks about his views on business, life, and philosophy. In these communications is introduced a clear and specific way of looking at how people should provide service to the business and to society at large, and most importantly, how to manage this process as a leader.

To put it more precisely, Ren Zhengfei explains leadership as a "bucket of paste." In a company like Huawei, where there are more than 170,000 people at work, it is necessary that something bind them together, a kind of glue to combine the force of all the employees and to focus them on the company's primary goal: providing the best possible service to the customer. According to Ren Zhengfei, this kind of glue does not include trust, but is rather represented by a bucket of paste that contains philosophy, culture, and values. It is a highly complex mix of ideas, which is managed by a collection of systems. For that reason, Ren Zhengfei does not believe in immediately trusting people to pursue their jobs with almost no restrictions, but rather emphasizes the need of systems to control the self-interested side of humanity. His view of human nature is thus not that of good people who will do good things, but of people with weaknesses that need to be constrained and managed in the pursuit of their interests.

His perspective on humanity is based on developments that took place in the West versus China. Specifically, he admires the decision of the West to be free but with restrictions, which are needed because only the power of systems can control human greed and selfishness. As a matter of fact, being confronted with the depths of its Dark Ages, Europe started to put systems at the heart of its civilization. In China, during the glory years of the Tang and Song dynasties, the focus was to rely on the positive powers of human nature. The result of this difference in focus was that China declined and Europe became powerful. Ren Zhengfei believes that China's assumption that people are naturally good at birth was the wrong one. In his view, people are not born as good people and therefore need systems to enable them to achieve good things. He relates this to why we have education systems in place in our society—that is, it is clear that excellence cannot be achieved in our societies without education and that is why people receive education when they grow up. In a similar vein, we need systems to mentor and guide employees.

For Ren Zhengfei it is clear that initially, little trust can be given to people without creating a collection of systems in which they can develop. Therefore, trust in systems is the way to go, as it will help in creating value by ensuring that all the work needed to become successful is actually done. If systems are implemented efficiently, only then can a fertile ground be created for trust to ultimately emerge between people.

Ren Zhengfei's perspective on the meaning of leadership reflects that he has a clear—and sometimes even extreme—viewpoint on the world we all live in, but it is exactly that strong belief in his philosophy that makes him attractive (even charismatic) to people. And, despite the fact that many people have difficulties understanding the true impact of his words and thoughts, people also realize that the first one to criticize his ideas is Ren Zhengfei himself. Indeed, he is known to be strongheaded in developing his ideas and thoughts, but at the same time, he engages in much self-reflection and criticism, which leads him to be optimistic when the company is not doing well and stressing a sense of crisis when the company is doing well.

To make more sense of how a complex leader figure such as Ren Zhengfei, with some outspoken ideas, has led to the emergence of Huawei as a successful global company with Western values and a Chinese foundation, we bring together the different leadership dimensions and corresponding elements of his character in Table 9.1. The different elements are discussed in light of three important dimensions that characterize the personality of Ren Zhengfei: strengths, area of focus, and heart (see Table 9.1).

Table 9.1. The Three Leadership Dimensions of Ren Zhengfei

Leadership Dimensions		
Strengths	Area of focus	Heart
Adventure and risk	One focus	Inspiration
Strong will	Service	Humility
Purpose	Power of thinking	
Adaptive vision	Controlled democracy	
Coexistence		

The Individual Leadership of Ren Zhengfei

Strengths

The most effective leaders are those who know what they are able to do and, because of this knowledge, walk in front, are curious, are not risk-averse, and know when to execute and when to wait. As Ken Kesey, an American novelist, once said: "You don't lead by pointing and telling people some place to go. You lead by going to that place and making a case." Making use of your strengths, which also implies at the same time knowing your weaknesses, creates the kind of leadership that is able to transfer strength to others so that they know how to operate. Looking at who Ren Zhengfei is as a person, it quickly becomes clear that he possess certain characteristics that enable him to empower others in such ways.

Adventure and risk: For one thing, he has a desire for adventures and does not shy away from taking risks. Being a romantic soul, Ren Zhengfei adopts a positive view while learning from the world around him. This shows that he has an adventurous way of looking at things. For example, his early learnings were very much influenced by looking outside of China to see and learn what the future could bring. In fact, as a Chinese entrepreneur, he has always been fascinated by Western influences. Hong Kong is actually one of his favorite cities because in his view, it is the place where East and West meet. Ren Zhengfei appreciates the fact that Hong Kong was a colony for over 100 years and the Chinese people of Hong Kong used those years to comprehend Western culture and digest it into a Hong Kong culture.

Of course, his fascination with the West did not stop with Hong Kong. A memory that has always stayed with Ren Zhengfei is that at the beginning of the reform[1], the Chinese people did not really know what the world was like. As a result, this situation made it difficult to compare and decide what was normal and what was not. One of the stories he holds dear to his heart is the idea that when he was young, they did not have much to eat, so they were convinced that everyone in the world was hungry like they were, or was even worse off. When the country opened up, it was suddenly possible to go overseas, and at the end of the 1980s, he found himself in a situation in the United States where he and his companions were too timid to eat the bread rolls in a restaurant. These bread rolls were on the table, but they had not ordered them. Why were they there? They wondered whether they would have to pay if they ate them. Because in China, if you wanted to buy bread rolls, you needed coupons as well as money. Nevertheless, Ren Zhengfei tried

---

[1] The Chinese economic reform starting in 1978; literally "the Reform and Opening up."

a few rolls, and ended up eating the whole basket. To their great surprise, the waiter brought more. Even more of a surprise, at the end of the meal, the rolls did not have to be paid for. They found out that the same practice was found in other restaurants as well. This personal experience made Ren Zhengfei realize that the world had many ideas and other ways of business to offer, which could be regarded as important inputs to his efforts at thinking out of the box while building Huawei over the years.

Transforming Huawei into the company it has become now clearly needed a sense of adventure and the willingness to take risks. His fascination with the West helped in pursuing this adventurous journey and, in combination with his engineering studies and time with the PLA, where he worked on military technologies, made him a leader willing to face the consequences of his sometimes risky actions. Take for example the year of 2001, when Huawei was almost acquired by Motorola. When the deal was almost done, Motorola appointed a new chairman and this chairman eventually rejected the last offer. A setback like this forced Huawei to take risks in order to survive. These situations also make clear that the growth of Huawei into the status it enjoys nowadays has not been a straight line of success. Because Ren Zhengfei takes risks regularly, the company has also failed several times, especially when exploring the global market. All these experiences, however, only made Huawei stronger and Ren Zhengfei more convinced that more effort, dedication, and passion would eventually help the company to achieve its goals. As one Huawei executive said when talking about Ren Zhengfei's way of making decisions, "even if we only have 30 percent confidence in the decision, we will still take the risks, because we believe that the other 70 percent will come from our willingness to sacrifice and show dedication at every level of the company."

It is an attitude that eventually paid off, because in 2008, when the financial crisis happened, many Western companies were forced to restrict costs. At that point, the many sacrifices made and risks taken in the past led Huawei to be very aggressive in recruiting people. There is some irony to this story: If the financial crisis had not happened, Huawei would maybe have had to wait several more years to grow. So, because of many global and local challenges, the style of doing business at Huawei has always been relatively aggressive in the last 20 years. Even more so, it was accepted for a long time that you need to be aggressive to survive and move on. Huawei has now arrived at a point in time where it is turning from a follower into a leader, which is yet another enormous challenge, but now one where it needs to find a compromise between taking less risk and taking on new adventures.

Determination: Being as motivated as Ren Zhengfei was in making Huawei a global player and now is in ensuring that Huawei has a sustainable future indicates that he has a strong sense of determination. To persevere is a value that has been there in him since childhood. He was born in 1944 in Southwest China's Guizhou Province, which was one of the poorest regions in China, a childhood experience that taught him not to take anything for granted. He had six brothers and sisters and the circumstances made it difficult for a family of nine to survive. Poverty and hunger—they even had to eat grass at one point—were part of his early memories and serve as the foundation of his displays of perseverance and dedication.

This characteristic of determination plays an important role in how Ren Zhengfei's leadership affected the development and growth of Huawei. In the early years, Huawei lacked funding, talent, technology, and management expertise, which was a huge disadvantage for this first generation of Chinese entrepreneurs in their battles with Western companies and SOEs. One thing that allowed the company to survive many setbacks is his strong belief that he was able to transfer to everyone else within Huawei. In fact, Ren Zhengfei believes that without strong beliefs, Huawei would have collapsed long ago. In 1999, Ren Zhengfei visited the Voortrekker Monument in Johannesburg, which honors the Dutch immigrants who spent nine years moving from Cape Colony to the continent's interior in the 19th century. After leaving the monument, he cried for hours. After seeing how hard the lives of these Dutch immigrants had been, he thought of how hard Huawei had suffered during its first 10 years. Since then, in his speeches, he has consistently emphasized that Huawei must survive. On one occasion, someone asked him what Huawei's most basic goal was. He replied: "survival." The person then asked what Huawei's ultimate goal was. Ren Zhengfei replied that it was also survival. If not for the strong will and beliefs of Huawei's founders and leaders, the company might have failed long ago.

Great leadership lies also in the ability to transfer such strong beliefs to the habits and behaviors of those who are being led, and this is something that Ren Zhengfei, together with his top executives, keeps doing every day. Due to the growth and status of Huawei nowadays, the company's biggest problem has been resisting temptation. It battles daily with the lure of opportunities. The senior leadership does not have the time and energy to check every budget, every year; departmental budgets are approved at each different level of management. Some of the younger managers in the company are not always content to keep doing the same thing, because breakthroughs are hard to achieve and require much experimentation. As a consequence, they may be tempted to shift their budgets and spend them on areas outside of the core business, which may lead to wasting billions of yuan. If it were not for the persistence of the senior leadership—their perseverance in focusing on one single area—Huawei may well have long ago dissolved in a puddle of diversity.

Value-driven leadership: Excellent leaders are able to communicate to their employees the purpose they and the company should pursue. The core of Ren Zhengfei's leadership is the undeniable fact that he is very clear about the purpose of Huawei, which is helping customers to realize their dreams. The core value of Huawei thus lies in making sure that customers come first. Ren Zhengfei truly lives this value and pursues it with unlimited energy, making Huawei his life's mission. In a way obsessed with doing good for the customer, Zhengfei is relentless in communicating this purpose to his employees by means of stories and in convincing them that they all should be dedicated to pursuing the mission of the company: "to connect people via communication."

To achieve this, it is agreed that Huawei should always listen to customer's needs and expectations and this input should fuel the whole enterprise. Important to realize is that such a value-driven focus implies that product development is not simply based on a reactive strategy towards what the competitors are doing, but rather on a belief that transformations happen with close collaboration between the developer and the buyer. This means not simply producing to produce, but rather production based on the recognition of true needs. In this way, Huawei embraces the idea that real innovation happens because a shared interest in values is created.

Adaptive vision: Fuelled by his famous sense of passion, Ren Zhengfei works hard to translate the purpose of the company into a vision that will help Huawei achieve a leading status globally. While pursuing this vision, he has continuously proven his ability to design strategies that enable him to adapt his vision to the challenges that the company faces. Important in this whole management process is that his use of an adaptive vision never compromises the purpose and values of the company. This ability derives from his proactive attitude. He is continuously focused on the future. As a matter of fact, one of the strengths that people usually refer to when talking about Huawei's founder is that he is always thinking about the kind of company he wants Huawei to be in the next 10 years. For example, Huawei plans the development of the company by decades whereas most of its competitors, such as Ericsson and Motorola, plan it by financial quarter or year.

His ability to critically reflect on past successes and, at the same time, identify the future challenges in the next decade makes Ren Zhengfei an impactful business leader in the eyes of many Chinese people. Indeed, as a Chinese saying goes: "Every generation produces its own great character" (*Jiangshan dai you rencai chu*) and "each exerts impact for hundreds of years" (*Ge ling fengsao shubainian*). True to this saying, Ren Zhengfei has proven that he had the influence and insights needed to design the most effective strategies to ensure that Huawei has grown into a leading company globally across three phases, each lasting about a decade. Interesting in this journey so far is that each phase is characterized by a specific focus and strategy.

The first phase ran from 1987 to 1997 and can be characterized as the years of chaos, in which Huawei was trying to survive the entrepreneurial stage. It was all about hard work to bring the company up to the level that it was able to provide higher-quality service. In the second phase, running from 1997 to 2007, Huawei became more structured by hiring the management consultant company IBM to implement (Western) management structures, as these were completely lacking at that time. In the words of Ren Zhengfei, "Chaos was removed and structure entered Huawei." The hiring of IBM also made it possible for Huawei to learn from the West, to allow the introduction of a more global vision. Ren Zhengfei was very clear about this ambition and demanded that all Huawei employees adopt the US way of looking at business as introduced by IBM. He made this persuasively clear by literally saying that it is necessary "to cut off your feet to wear the shoe" (*xuezueshilu*). In his view, in the second phase, it was necessary to wear the shoes of the Americans and if your feet would not fit, then you have to cut it. This statement was understood as underlining the fact that his global ambitions were there to stay, and grow. In the third phase, after 2007, the strategy was to simplify management and be creative with top talent to turn Huawei into an efficient and creative creator of customer's dreams. Ren Zhengfei was worried that the second phase created a kind of decision making that was less than efficient. Decisions were taken much slower than in the first—more chaotic—years of the company and there was a fear that the bravery and courage underlying the company's striving for innovation would disappear. For that reason, the third phase focused on simplifying management and creating conditions such that more creative chaos could be brought back into the structured management models.

Bringing opposites together, or coexistence: One defining characteristic of the Huawei culture is that it brings together opposing forces and tendencies. One set of opposing forces is the idea of "attack" versus "compromise", also referred to as the simultaneous tendency to cooperate and to compete. As we noted earlier, the first two decades of Huawei's existence were dominated by the force of attack in order to survive and become gradually a better service provider. At that time, Ren Zhengfei clearly used a competitive mindset as the driving force to grow the company. Essential in his idea of competition, however, was that it should be done with respect for Huawei's opponents.

Where does this idea of competing in a cooperative way come from? Ren Zhengfei is known to value very much the importance of history with regard to how well a company can fare. He has been quoted as saying that if there were no reform or opening up of China, then Huawei would not be here. He believes that there was not any lack of talent in the generations before his, it was only that they just did not live at the right moment in history. According to Ren Zhengfei, we have been given a great age to live in, with almost unlimited opportunities. He compares it to riding on a flying carpet: It is not us who are flying, it is the carpet carrying us upwards. In a similar vein, his interest in and studies of foreign history have influenced Huawei's strategy of balancing competition and cooperation. This specific company philosophy was inspired by the heroic tales of the Glorious Revolution that took place in England in 1688. This tale tells the story of how King James II of England was overthrown by a union led by William of Orange in 1688, which was called the Glorious Revolution. This revolution was also referred to as the Bloodless Revolution because the victory of William of Orange was achieved without bloodshed. This fact inspired him to embrace the idea that one can win and still be cooperative.

The idea of adopting a cooperative mindset in a competitive market is somewhat of a brand trademark nowadays and can be illustrated by Huawei's efforts in the United Kingdom. Huawei has put much effort into convincing the British government and the general public that it and its procedures can be trusted by (a) setting up the Cyber Security Evaluation Centre in Banbury to ensure the quality of their equipment and (b) cooperating with the Government Communications Headquarters (GCHQ), the United Kingdom's signal-intelligence agency, to ensure that the networking equipment and software is reliable and secure. In fact, the recent and growing success of Huawei in Europe can also be attributed partly— in addition to their service-oriented focus (see again the point of purpose-driven ambition as explained in this chapter)—to its philosophy of explicitly incorporating a strategy to develop cooperative relationships with competitors in the market. Indeed, EU officials initially wanted to investigate the anti-dumping act in relationship to Huawei's products. Despite this controversy, Huawei received support from both Ericsson and Nokia, stating that in their view Huawei was not dumping its products.

Areas of Focus

A key characteristic of wise leaders is that they are able to grasp the essence of things. As Ben Stein, an US writer, so nicely puts it, "The indispensable first step to getting the things you want out of life is this: decide what you want." Possessing this skill makes leaders effective as they have a purpose, a direction, and an adaptable attitude to achieve the goals they strive for. In this process, it is also important that those leaders are able to direct the attention of their employees to the outcomes that need to be created. They can do this by emphasizing the mission, vision, values, and strategic goals of their organization and at the same time building the capacity of their organizations to achieve them. This skill of focusing on what matters fits the image of Ren Zhengfei, as he has been referred to as someone who is able to communicate to others the essence of what they do and why they do it.

Focus on one thing to become better: According to Ren Zhengfei, one key aspect of why Huawei became the leading Chinese company is their strong focus on crucial and continuous investments in R&D to make their products better. In the 1980s, China lagged behind in telecommunication. There were more than 400 companies all over China and the profit growth was very high. As a result, most of these companies lacked the motivation to perform better because it was easy money. Such an attitude, however, makes you lazy and, in the long term, a weak competitor. It is essential that you keep focusing on one thing and become very good at it.

Ren Zhengfei is very much convinced of this philosophy of being focused on one thing. He once told people that water is soft, but in Germany, they use high-pressure water to cut steel plate. Air is also soft, but rocket engines can use that same air to propel a whole rocket. These examples make clear that you can generate much power and impact when you focus on a single point. In a similar vein, at Huawei over the last 28 years, employees have basically all been working on the same thing. This focus has led them to invest over CNY100 billion directly in a single area (CNY50 billion in R&D, CNY50–60 billion in marketing and services), and this investment has made Huawei a global leader in massive data transmission.

Service: Part of making the dreams of customers come alive is the idea that providing the best service possible is crucial to the success of the company. In fact, Ren Zhengfei has always made clear that serving customers is the only reason Huawei exists. In his view, customers are the soul of the company and thus should be the focus of the company.

In the early years of Huawei, Ren Zhengfei was very much aware that its products did not meet the standards of those of its competitors. For that reason, he looked for alternative ways to attract customers. In his view, this could only be achieved by providing excellent services. For example, Huawei's equipment broke down frequently, so Huawei's technical engineers went to the customers' equipment rooms to repair the equipment at night when the equipment was not being used. Huawei responded to customers 24/7. This practice differed from the ones used by Western companies, which had advanced technologies and equipment, but ignored their services sector. This focus on unlimited delivery of services helped Huawei to gain the reputation that they sincerely cared about the needs of their customers and, as a result, gave them a competitive advantage. For example, in the desert and rural areas in China, rats were a plague for telecom connections to customers. The multinationals that were present in China at that time did not consider this to be their problem, but rather, that of the customer. Huawei's competitors assumed that they only had to provide the technology to the customer. Huawei, in contrast, decided to provide services to their customers. They viewed the rat problem as one the company had the responsibility to solve. An interesting consequence was that, because of this purpose-driven strategy, they acquired much experience in developing robust equipment and materials, which helped them later on to win several big business accounts in the Middle East.

Power of thinking: A common theme underlying everything that Ren Zhengfei undertakes as a leader is his ability to reflect, think, and act. He is frequently quoted as saying that the most important thing to value is the power to think. This belief is built on the idea that innovation without a solid academic foundation is never going to become big business. It is just messing with the details. Back in the days of the war of resistance against Japan, there was a saying in China: "Across the whole North China plain, not one corner is quiet enough to set up a writing desk." Ren Zhengfei feels that the Age of the Internet has created a generation with inflated ideas among China's young people and therefore fears that serious scientific research is lacking these days. The only way to achieve great leadership is therefore to do some hard learning.

Important to the development and growth of Huawei is that Ren Zhengfei has always pushed the idea of seeing the power of thinking not as an important personal characteristic but, primarily, as an essential part of the company's culture. According to him, Huawei should be focused on building a company where people's minds are the main asset and resource to rely on. The importance of thinking, in his view, is that it provides the skills to connect the dots needed to work with an agile vision and strategy. Ren Zhengfei indeed works hard to always keep clear to himself a kind of meta view that enables him to make informed strategy decisions, because he is convinced that failure will come one day, so the company therefore must be prepared.

Interestingly, such a strategy suggests that a power-of-thinking attitude aligns with a shared-learning orientation. This orientation is very visible in Huawei since it has invested considerably in creating a learning-driven culture where the power of the mind is visible. As we explained earlier, often references to past historic events are used to shape actions and beliefs towards the future. Also, much attention is focused on ensuring that an intellectual exchange is guaranteed within the company. For example, executives are encouraged to read both specialized and non-specialized books to foster an intellectual climate. Although none of the board members has studied overseas—they are all educated in China—they all are committed to forcing themselves to be open-minded enough to absorb as many new insights as possible. Ren Zhengfei is careful to make sure that he sets the norm for this intellectual enterprise in a variety of ways: (1) One of his main hobbies is reading and Ren Zhengfei is known for, when he finishes a page, tearing out that page so he will have to remember what he read and (2) for engaging in extensive interactions and talks with politicians, scholars, and representatives of religion to acquire the best worldwide perspective possible and learn what is valuable for the strategy Huawei should pursue.

Controlled democracy: In China, leadership has traditionally been grounded in a hierarchical top-down management system. Huawei is, to a large extent, no stranger to this system. The leadership of Ren Zhengfei, however, does differ from this rather "controlling only" style in important ways.

On one hand, he possesses the characteristics of a controlling leader who makes all the decisions. As can be expected of a man with an army background, he is known to be intense and tough and never loses control. In the early years of Huawei, this toughness was symbolized by his focus on fighting and surviving as a primary strategy for his people. The slogan at that time was: "We shall drink to our heart's content to celebrate our success" (*sheng ze ju bei xiang qing*), "but if we should fail, let's fight to our utmost until we all die" (*bai ze pin si xiang jiu*).

On the other hand, Ren Zhengfei is also known to give his people much freedom when it comes to how decisions are executed. In Huawei's early years, Ren Zhengfei retained authority on major decisions such as corporate development strategies and culture building, but fully empowered employees when it came to R&D, manager appointment, compensation and benefits allocation, and other areas. This practice allowed managers to become more active and creative, but also created chaos. After learning from the West for almost 20 years, Huawei has established an increasingly standardized and institutional decision-making system. Collective decision making ensured that Huawei made fewer mistakes and absorbed collective wisdom, but it also caused rigidity. Therefore, Ren Zhengfei acted more like a catfish in the high-level decision-making process, and has constantly created imbalance to inspire passion across the organization.

Today, Huawei has developed a decision-making system with limited democracy and appropriate centralization. Such a system prevents the company from collapsing due to reliance on a single leader, raises efficiency, and avoids inaction due to over-democratization. This system can best be understood by comparing it to a fireplace. Within a controlled space (the fireplace), people can be free to a very large extent (as the fire is within the fireplace). This idea is closely related to Western management thinking in which the relationship of a person to organization or a person to a company is considered to be contractual. It gives you rights, but at the same time, these rights must be bounded. If your rights are unlimited, Ren Zhengfei believes that it will destroy the organization and other peoples' rights along with it.

Heart

Leading with the heart and trying to make a difference in people's lives makes for great leadership. As John C. Maxwell, an US author, once noted, "People do not care how much you know until they know how much you care." Having a willing heart and a positive attitude allows people to connect to you and feel supported in the decisions they make. In this process, it is important that leading with the heart is motivated by your true and authentic self because it will ensure that you make decisions based on your own values. Here, rhetoric plays an important role as motivating people by means of your values requires you to be able to touch the people's hearts and minds. Exactly this aspect of effective communication and being authentic represents the heart of Ren Zhengfei's leadership.

Inspiration and storytelling: Creating a committed workforce in service of the company's purpose requires that people are inspired. One of the qualities that has consistently been attributed to Ren Zhengfei is his ability to inspire others. It should be no surprise that, for that reason, Ren Zhengfei is a big fan of the practice of storytelling. Storytelling is at the heart of every business success and it can take place at any level inside and outside the company. It enables business leaders to connect the story of the "bigger picture" to the company's vision and this is communicated to employees, clients, and stakeholders that determine the deals that are made.

In the early years of Huawei (phase 1), he believed that in 20 years, Huawei would have a third of the market share in the world, despite the fact that at that time Huawei only employed about 200 employees. Many of them thought at that time that he was crazy. Nevertheless, Ren Zhengfei remained persistent in talking about his ambitions on all kind of occasions. Famous is the story where Ren Zhengfei, in Huawei's fifth year of existence, suddenly rushed out of the kitchen when cooking for Huawei employees abroad—his hobby is cooking—and announced, "Huawei will be a top-three player in the global communications market 20 years from now!"

Underlying his focus on storytelling is his strategy to energize people and introduce a sense of vitality to the projects that need to be undertaken. Especially in the entrepreneurial years of the company, he used this strategy to be both a strong mentor and a leader for his employees. As a leader, he was used to delivering a vision, and as a mentor, he was guiding his employees towards their goal. For example, as the products of Huawei in their pioneering years were not yet well-developed, he visited many R&D offices abroad. During one of these visits, in 1997, he visited Bell Labs in the United States. The story goes that Ren Zhengfei was so moved by the work done in those labs that he cried. Back home in Shenzhen, Ren Zhengfei told everyone that the passion that he felt for that lab was equal to love. Bringing them this emotion-laden message was meant to motivate his own people and eventually to make his R&D people believe that they could one day be better than the researchers at Bell Labs. Inspirational leadership, as Napoleon Bonaparte once said, is a matter of dealing in hope—something Ren Zhengfei has as a trademark in his leadership style.

Humble leadership: In his pursuit of the Huawei dream, Ren Zhengfei seems to know his own limitations and does not portray himself as the ultimate know-it-all leader. He always emphasizes his belief in the value of talking from the core—your values that matter—to do "good" for the organization. His philosophy in that way is best served by displaying humble leadership. In the early years of Huawei, he frequently apologized to clients for the poor quality that it was providing at that time and tried to accommodate their needs as much as possible. In a sense, it was Ren Zhengfei's determination and focus on improving quality all the time that ensured that Huawei gained market share by being cheaper than international providers but quickly becoming better in quality than its Chinese competitors.

Although his inspirational leadership style pushes forward the company and guides it through transformational phases, he is always quick to add that he may not be that good a leader as many would like to believe. He is humble and very careful not to feed the myth of his leadership and rather likes the company's track record to speak for itself. For that reason, he has always had little need to show himself to the larger public. This attitude has led some people to argue that his behavior is suspicious as he seems to be hiding things, whereas others have argued that he is not interested in promoting his personal reputation but more the image of Huawei.

Being humble indicates that a leader knows both his weaknesses and his strengths and makes use of that knowledge to create value. In a similar vein, since the beginning of Huawei, Ren Zhengfei has embraced the reality that he is not a technical expert and at the same time stressed that this was not a weakness. Instead, he believed it was a strength. He believed that the combination of his skill at organizing a company and the IT background of his executives and employees would create wonders. He is praised for his desire to hire better people in the areas he is weak in, with the ultimate goal to improve the quality of Huawei's products and services.

Shared Leadership at Huawei

A final defining element that shaped the blueprint of Huawei and that can be seen as emerging from the combination of Ren Zhengfei's strength, focus, and heart is the fact that leadership is recognized as a collective effort. As the famous 19th-century US philanthropist Andrew Carnegie, noted, "No man will make a great leader who wants to do it all himself, or to get all the credit for doing it." In a similar vein, at Huawei the idea exists that the wise leadership exhibited by Ren Zhengfei builds on the idea that a leader's primary task is to build the context for the others to realize their potential.

The first illustration of this philosophy is the company's focus on sharing the rewards that are earned—they are after all the result of collective successes. For this reason, Ren Zhengfei only holds 1.4 percent of the company's total share capital whereas 82,471 employees hold the rest of the shares (stated in Huawei's 2014 annual report, published December 31, 2014). This performance incentive system ensures that people are not only motivated to work hard towards collective success, but probably even more importantly, it ensures that Huawei really is an employee-owned company.

The idea of truly committing to be an employee-owned company also shows how Huawei has taken the lead when it comes to being innovative in the field of organizational leadership. Indeed, after the financial crisis in 2008, the world's attention increasingly turned to the issue of wealth distribution. At a societal level, discussions emerged recently about the ever-widening gap between the super wealthy (the so-called 1 percent) and the rest of the world population, and Western governments particularly began thinking about ways to tax in fairer ways. At the same time, in the corporate world also, the issue of wealth distribution has become an important topic. For example, a report by the American Federation of Labor and Congress of Industrial Organizations (AFL-CIO) revealed that in 2014, top CEOs earned 373 times more than the average US worker. Wealth distribution has also been linked explicitly to employees' levels of productivity. Research shows that perceptions and experiences of significant wealth inequalities are regarded today as damaging productivity and feelings of ownership in employees. This assumption has even reached the minds of politicians running for the 2016 US presidency. For example, Hillary Clinton has argued that the time is ripe for companies to start offering profit-sharing plans to their employees. Clinton has noted that in order to increase productivity, wages have to go up.

It is clear that to achieve such an outcome, it is necessary that companies allow employees to benefit from the profits that companies make or, in other words, implement a structure where employees own (at least partly) the company. Although much debate exists on this topic, it is very rare that a global business leader actually embraces such an innovative way of leading an organization as is the case with Huawei. However, knowing the founder's sense of adventure and risk, it may also come as no surprise that Huawei has a culture where an employee stock ownership plan (ESOP) is possible. The structure of the ESOP is based on two important premises. The first premise is built on the Confucian values of equality and harmony, and underlies the ambition of Ren Zhengfei to ensure that the gap in wealth between employees—and by extension, across society—should not be too big. To achieve this, qualified employees are provided *equal access* to the ESOP, allowing them to engage in projects that could earn them higher salaries, bonuses, and promotions. Although Huawei is big on creating equal access to those opportunities, they do not employ a view that promotes equality in the distribution of outcomes. At Huawei, the second premise is built on the idea of *equity* and not equality. The harder you work, the more you can earn. It is important to note here that working overtime is only rewarded extra if the work directly addresses the needs of its customers. Projects in overtime that do not reveal direct positive consequences for the welfare of the customers are not paid out extra.

This idea of equity in how the final outcomes are distributed is a crucial determinant of motivating employees to keep putting in their best work, as was recently demonstrated by the decision of Dan Price, CEO and majority owner of Gravity Payments, to increase the minimum salary for all his employees to US$70,000. Although initially everyone endorsed this example of enlightened capitalism, several of their most skilled employees left the company as they did not consider this policy to be fair—they resented the fact that the highest raises seemed go to the ones who showed the least skills. Huawei wants to avoid exactly this scenario, as it wants to create the right circumstances in which its employees do show their abilities. Its employee-sharing scheme is not only focused on allowing employees to earn more and control wealth gaps, but also to allocate power to those who show strong skills. Ren Zhengfei has always endorsed the philosophy to respect human nature and work with it accordingly. In his view, people do care about belonging to and being proud of a collective—which Huawei is—but also have the desire to differentiate themselves from others by means of power. Looking at it from this perspective, the foundations of Huawei's ESOP can satisfy both human needs.

A second illustration that Huawei puts shared leadership into practice concerns the development of Ren Zhengfei's thinking that emphasizes that for the company to prosper in a sustainable, long-term manner, he has to take himself out of the equation. It is too dangerous that one person does it all. Too much power for one person is not good. The same idea is applied to how markets should work. Ren Zhengfei repeatedly stresses that to have the best market available, you need competitors. It is quite easy: Relationships matter. Ren Zhengfei is quoted often to have said: "I do not know anything about technology, but I can bring people together to work for the collective." This statement signals a strong belief that there is infinite strength in organization and collective efforts. Great things, in his view, can only be realized if everyone is aware that as individuals they are not that significant. It is all about working together. He claims that when he founded Huawei, he no longer acted as a technical expert, but became an organizer.

This philosophy clearly influences how management is structured at Huawei. And, as we have noted several times, Ren Zhengfei is fascinated by many Western developments of the past, but nevertheless, he does point out that the Western management model has a flaw: It loads all of the company's hopes on to a single person. But there is no guarantee that illness or an accident will not strike. Huawei does not depend on any single person, so they do not have to worry about the consequences of any accident. Ren Zhengfei himself has illustrated this idea with the following story: Everyone agrees that George Washington was a great leader. He oversaw the drafting of the US Constitution, which has been a defining influence on the United States for more than two centuries. But for 109 days at the Constitutional Convention, all he did was bang his gavel at the right time in the morning and bang it again in the evening. When they came to vote, three delegates to the convention voted against the Constitution and 47 voted to approve. So for a company to be long-lived, it should be led by collective wisdom.

This idea of collective wisdom is also behind Huawei's use of the rotating CEO platform. Huawei has decided to counteract the possible negative consequences of having only one leader (CEO) by introducing a system where the acting CEO changes regularly as well. Ren Zhengfei's leadership has created diverse views on how the company should operate but again in a controlled manner: No CEO will rule too long (6 months). Adopt a balanced approach, where you have an eye on the future, a focus in the present, and a learning orientation towards the past. This is also what Ren Zhengfei advises young entrepreneurs today.

The idea of rotating CEOs has been based on examples from Western leadership models and the animal kingdom. The rotating platform can be seen as modelling the Constitutional monarchy in the United Kingdom. The chairwoman of Huawei's board is like the British queen: She has no actual power. Power is held by the parliament, and the parliament follows majority rule. Even if an extremist leader does appear, he needs the support of more than half of the members of parliament, or he is powerless. Furthermore, the idea of having a fixed time period (6 months) to rotate the sitting CEO out is further inspired by the four-year rotating system that is used for the presidential elections in the United States.

Within Huawei, the belief also exists that we need to respect and learn from nature itself. Insights from the animal kingdom as such provided input to develop the idea of the rotating CEO system. The first insight relates to how herds of buffalos live and function together and is inspired by a book on new leadership called *Flight of the Buffalo* by James Belasco and Ralph Stayer. In this book, it is explained that herds of buffaloes follow their chief buffalo and if this chief is killed, the herd will end up in chaos and die as well. According to Belasco and Stayer, this story represents the traditional leadership style where one leader gives direction and therefore becomes indispensable, creating a situation where followers will not know what to do if the leader is not present anymore. In a similar vein, Ren Zhengfei does not want Huawei to be clueless about its own business direction if he—as the only CEO—passes away. To avoid this situation, a second insight story is relevant. This story is about how ducks fly when travelling in groups. Ducks tend to fly in a "V" formation and what is specific to their behaviour is that the leader of the group changes regularly, allowing others to rest.

The Future Leadership

All of the above clearly points out that the roots of Huawei lie with an individual with unique capabilities, who founded a company but then became part of that company, which keeps growing with respect for its own history, its embeddedness in both the Chinese and Western cultures, and its desire after so many years of being a follower to become a leader. It all suggests that leadership at Huawei is highly valued but always in the service of the company's interest that is shared by all of its employees.

As a consequence, speculations and worries about what will happen at Huawei when their founding leader is not there are very much alive. And those concerns identify two important leadership challenges. The first one is underlined by the observation that maybe the ultimate goal of the rotating CEO system is to identify the successor of Ren Zhengfei. As we mentioned earlier, the business model at Huawei is based on values that Ren Zhengfei has made important since the inception of the company— serving the needs of customers by being sincere, persistent, and innovative in making the dreams of customers come true. A potential worry therefore is that if Ren Zhengfei is not part of Huawei any more, whether it will be able to remain such a strong, value-driven organization. For that reason, the Western media especially has suggested that this CEO rotating system is used as a kind of entrance exam to determine who will become the next leader of Huawei and thus succeed Ren Zhengfei. However, Huawei has become more than an individual, a social identity that is shared by those populating Huawei and therefore the company does not seem to be looking for a replacement for Ren Zhengfei at all. The most likely outcome will be that in the future, many more rotations will take place, making many more excellent talents come to the surface, but importantly, no dictators will be born. The rotating system is a restraint on the individual and, as the cycle repeats, brilliant people will have many chances to demonstrate their quality.

The second future-leadership challenge concerns the continuation of the transition process from being a follower to a leader in the telecom industry. Huawei is a member of 146 organizations; they hold chairs, deputy chairs, committee memberships and they have started to play a role in global leadership. In addition to this, they give funding to universities in Europe, Japan, and Russia; and it is expected that in the future, they will also give significant funding to US universities. So, it is clear that Huawei has closed the gap with Western companies and to some extent even overtaken them. As such, the follower has become a leader and now has to figure out its own path.

One crucial direction that Huawei is emphasizing and that should be the backbone of any innovative project in the future is the transformation of the company into a truly learning organization, with shared leadership and management through values. A learning organization is characterized by the presence of a learning climate that is participative in nature. In this climate, senior managers and junior employees work and experiment together to try out new ideas and failure is allowed in the pursuit of growth and progress. Across different levels within the company, everyone is allowed to participate in the value-creation process. And precisely because of this reason, it is important that the values that make the company are widely recognized and supported. Huawei is paving the way for such a change, so that the change is embraced, allowing innovation to follow. There is little concern that this important aim will not be achieved because, after all, the proverb used in Huawei is: "The only thing unchanged is change."

Future Shock and China's Opportunities

Addendum

Part One

In 1970, Alvin Toffler, US writer and futurist, published a book called *Future Shock*. Ten years later, Toffler released a more influential work, *The Third Wave*. Looking back, for the past 30 years, mankind has been the living proof of this scholar's predictions for the future: the computer revolution, the Internet, open universities, increasingly deconstructed organizations, globalization....

It was in 1974 that the 30-year-old Ren Zhengfei came into contact with computers for the first time. He had listened to an academic report by Wu Jikang, a Chinese scholar who had just returned from his visit to the United States, on what computers were and their potential applications in the future. "I didn't understand a single thing he said throughout the entire lecture," Ren Zhengfei recalled, but that lecture had changed young Ren Zhengfei's life forever.

In 1979, the Third Plenary Session of the 11th Central Committee of the CPC removed historic restraints on commerce, and billions of Chinese people charged forth onto the developmental road of no return. The next three decades were filled with all manner of desire, passion, innovation, risk, and impulsive behavior, accompanied by anxiety, pain, hardship, sweat, tears, and countless sacrifice. While the fate of numerous individuals and families had changed, this ancient nation in the East also reemerged as a world power.

Destiny favored China. China's reform and opening up was marching to the same drum as the global technological revolution that began in the 1980s. The theory of international division of labor, proposed by a group of US strategists, had provided China with massive development opportunities. Coveting the immense Chinese market with its population of 1.3 billion people, already-developed countries like Japan, the United States, and those in Europe began transferring capital, equipment, technology, and management. Although China has paid a huge price for environmental damage and resource exploitation over the past few decades, we have to admit that the Chinese nation has at last arrived at the cusp of prosperity. There is still a gap between us and the developed countries, and many deep conflicts and problems still exist; China is nevertheless a far cry from what it used to be in terms of international status, economic power, and strategic confidence.

Now, 30 years later, mankind is standing on the threshold of another major turning point. The computer-centered technological revolution that occurred 30 years ago was built on traditional industrial society, or even on agrarian civilization for some countries, but this new era of technological innovation has some very essential differences. It doesn't matter if you're in developed regions such as the Americas, Europe, and Japan or in less-developed countries like China or developing countries in Africa; all of mankind has been profoundly and extensively swept up in the waves of globalization and information technology. The fundamental changes introduced by the Internet, in terms of knowledge and technology sharing, interpersonal interaction, and methods of trading commodities and capital, have caused the world to become flat. Accordingly, the Internet has granted individuals and countries, regardless of ethnical, cultural, and political background, the opportunity to participate in the course of globalization.

In the 1980s, when US strategists put forward the idea of international division of labor, they neglected a critical blank spot—that is, although the Internet was born in the United States, it put an end to the monopoly that Western countries had on the high-end industrial value chain. Thirty years ago, it would have never occurred to a select portion of arrogant and over-optimistic strategists that some of the developing countries out there, of which China is a prime example, would be able to keep pace with and eventually stand at the forefront of information technology and globalization and would end up nurturing a group of world-class, high-tech manufacturing companies such as Huawei, Lenovo, and Haier as well as Internet companies such as Tencent, Baidu, and Alibaba.

What will society be like in the next 20 or 30 years? In international forums on technology and thought exchange that involve China, like the World Economic Forum, tens of thousands of politicians, strategists, entrepreneurs, scientists, and scholars from the United States, Europe, China, and other countries have put immense thought into sketching out mankind's next steps. On the Internet, tens of millions of ordinary people, particularly young people, are also following and discussing the technological singularity that might possibly occur in the next two to three decades.

Eric Schmidt, executive chairman of Google, predicted that the Internet will be replaced by the emergence of the Internet of Things. Industry 4.0, led and promoted by German Chancellor Angela Merkel beginning in 2013, has been elevated to the level of national strategy in Germany in less than three years and has been widely acknowledged in industrial, academic, and research circles around the world. Industry 4.0 follows a bottom-up pattern and is intended to improve the competitiveness of German industry so that it can lead the new round of the Industrial Revolution. The United States, however, applies a top-down approach, leveraging the Internet to make industrial production systems (machines, data, and people) into smart systems. US industrial circles regard the "industrial Internet" as the third industrial revolution after the first Industrial Revolution and the Internet. This new revolution is led by giant corporations like GE, Cisco, and Intel, and is funded by the US government.

Meanwhile, material technology is making explosive breakthroughs. The era of graphene has already begun while the era of silicon is coming to an end; genetic research-based biological technology is also pushing humanity towards a new tipping point. The integration of information technology, new materials, and gene technology will lead mankind from several thousand years of traditional society into a brand-new era that is both exciting and terrifying.

Moore's Law is the collective obsession of scientific and technological elites. When a group of strategic "madmen" across the world look to the skies, imagining and designing the theoretical framework, ideological structure, and systematic architecture of future society, groups of scientists, engineers, and entrepreneurs, driven by capital and a handful of talented people, are hastening their chariots to close in on its blueprints.

Futurists such as Alvin Toffler are lucky to live in this fast-changing era, where they can stand aside and watch as most of their predictions come true.

Part Two

The US political scientist Francis Fukuyama once argued that democracy, free market capitalism, and autonomous lifestyles in the West might very well signal the end of mankind's sociocultural evolution. He enjoyed his 15 minutes of fame with the book *The End of History and the Last Man*. However, when his new book, *The Origins of Political Order*, was published 25 years later, he apparently no longer believed that the Western system is a one-size-fits-all panacea for all countries. A large part of the book is devoted to his reflections on the defects of the democratic and social-welfare systems in the United States. On the title page, there is a quote from Alexander Hamilton: "Energy in the executive is the leading character in the definition of good government. […] A feeble execution is but another phrase for a bad execution; and a government ill executed, whatever may be its theory, must, in practice, be a bad government." (Alexander Hamilton, *Federalist No. 70*)

Over the course of 30 years, at a stacking velocity, China has gone through the same process of industrialization that took the West 200 years to complete. This is closely linked with the government having identified the right direction and the sheer amount of energy it put into execution. Undoubtedly, many social and economic problems have piled up throughout this process of rapid growth, but at the same time, by relying on the country's and the Party's ability to mobilize society, the tradition of self-criticism, and mechanisms for correction, China is accelerating its transformation, though with difficulty, towards a market-oriented society that is governed by law. As a British observer has pointed out, the "Two Decisions" approved by the third and fourth plenary sessions of the 18th Central Committee of the CPC cover such a wide range of reform and were released with such speed that something similar would be completely unfathomable to achieve with decision-making mechanisms in the West.

Companies are one of a country's core competitive elements. A nation will prosper only when its companies grow strong as individual entities, in groups, and as complete industrial value chains. The national wealth accumulated by tens of thousands of companies in the past three decades has enabled China to have a say in world politics, economics, military, and diplomacy. However, all of this was predicated on the fact that economic reform unleashed the largest dividends in Chinese history.

However, we must soberly recognize that, after 30 years of catching up, although China has already approached the starting line for the next major paradigm shift, the internal and external environment for development has drastically changed. The traditional model of economic growth is no longer sustainable; the manufacturing sector is feeling the squeeze of the Internet and many other factors; and short-term profits seem to have become the main pursuit of entrepreneurial groups. To some extent, the challenges outweigh the opportunities for pretty much all enterprises in China.

The shocks that await us over the next 20–30 years are clear to see. Amid competition between the United States, Europe, and China, institutional fatigue has begun to seep into the democratic and social welfare systems in the West. China's problem, however, lies in the dramatic decrease of its reform dividends. It will take new and more deeply seated reform to lead businesses, institutions, and individuals from all walks of life out of the fatigue that plagues the nation.

The "Two Decisions" were made at the right time and are expected to unleash the country's full potential on an ongoing basis. If no "subversive mistakes" (President Xi's words) are made in the deep reforms to come, China will have a chance to stand out in its competition with the United States and Europe.

Further marketization of the economic sector will spur the latent vitality in all aspects of society to reach its maximum capacity and will facilitate the most effective mobilization and allocation of global resources like capital, talent, and technology. In particular, it will help greatly boost the confidence of entrepreneurs and start-ups. As Keynes stated, when economic prospects are uncertain, the "animal spirit" of investors and entrepreneurs, that is, the courage and adventurous spirit driven by optimism will largely determine how the economy pulls itself out of a downturn and heads towards prosperity once again.

One key factor in actualizing law-based governance is the protection of property rights and intellectual property rights (IPR). This is the cornerstone of the great world-class companies that were created in the United States and Europe. Chinese people have never lacked creativity. During the global technological revolution of the past 30 years, talented Chinese knowledge workers with international educations have become one of the main forces of innovation in academic institutions and large corporations in the United States. Over the past few years in Silicon Valley, the remarkable achievements of Chinese people who run their own start-ups or who are engaged in R&D work have continually demonstrated the creativity of our people. If the legal property and intellectual property rights of citizens and legal persons can be guaranteed against violation or infringement through effective execution of the law, it will attract more overseas students, engineers, scientists, and entrepreneurs to join in the competition between China and other developed countries for strategic high ground in the future.

In eras of peace, globalization can be likened to competition between marshlands, in that marshlands are low-lying regions into which water— or all factors of production, in this case—naturally flows. Thirty years ago, Deng Xiaoping, along with other great reformers and strategists in China, dug up a vast marshland of favorable policy in China, which unleashed our passion for creation on a large scale and caused floods of capital, technology, and talent to pour into China from all over the world. Thirty-odd years later in 2016, the new generation of Chinese leadership, with an even grander global vision and equally courageous top-level design, is very likely to generate a new piece of "marshland" in China, but this time it's going to have a strong sense of direction: that is, everything must occur within a law-based framework.

Part Three

China completed its "primitive accumulation," or formative years of hunting and gathering, within 30 years; but we have also paid—and are still paying—a huge price for the environmental pollution and resource exploitation brought about by a growth model based on low costs, low quality, and low prices. This has become a source of chronic pain throughout our country's economic and social development. We are gradually losing the strategic advantage we've gained over the past 30 years, or worse: This may cause us to be at a strategic disadvantage. All of China's 1.3 billion people need to quickly part ways with the "gobbling and gambling" model of development.

For China, the next two or three decades can either present a precious opportunity or spell the end of our prosperity. In this new wave of technological development that centers around artificial intelligence, competition between nations and businesses is essentially a competition for talent. The value of "population dividends" is a question that's worth discussing. In the last 30 years, China has taken full advantage of its abundant human resources, having seized the opportunities provided by major shifts in global industry to create the largest economic miracle in human history. However, today and in the future, our population will no longer be an advantage. In the next two or three decades, automation, informatization, and "going smart" will quickly become the new methods of generating and accumulating wealth. Traditional models of economic growth, consumption, service, and social governance, will also undergo fundamental changes. Therefore, China needs to transform itself from a populous country into a talent powerhouse.

Education is the foundation of a talent powerhouse. An increase of investment in education matters much more than an increase in population. In order to get the most out of an investment in education, you have to improve the social status as well as the welfare of elementary and secondary school teachers. History has proven time and time again that the modernization of any economy, be it the United States, Europe, Japan, Singapore, or Taiwan, is rooted in the modernization of its education system. The reform of our college entrance-examination system, which took place in the beginning of China's reform and opening up, sent millions of youth—whose education was put off during the Cultural Revolution—to universities. This helped to nurture a large number of talented people for the country across all different fields in the decades that followed. During the past two decades of urbanization and industrialization, tens of millions of young men and women left their rural hometowns in the hopes of changing their lives and the lives of their families and, at the same time, indirectly contributing to the development and prosperity of the country. However, the tremendous, invisible price of this mass exodus of labor was this: In the rural areas of the western and mid-western regions in China, a large number of children have been left behind without their parents, and their education has been neglected.

The restoration of education in China's rural areas is critical to the nation's future. "To cultivate a child is to eliminate a criminal" was an educational concept once advocated by the Japanese government. Similarly, if the majority of children in rural regions can receive quality education, both in terms of knowledge and proper citizenship, we can place ourselves in an invincible position for the next round of technological revolution. Currently, many scientists, government officials, scholars, and entrepreneurs in China either grew up in rural areas or have worked there for many years.

Of course, China's education should also place particular emphasis on developing vocational education throughout the country—for example, the cultivation of skilled workers, farmers, and village doctors equipped with modern technical capabilities. Vocational education in Germany, Japan, and Switzerland can provide us with many useful references. What China lacks the most is a spirit of craftsmanship. Merely advocating a culture of disruption won't help the economy shift gears from low costs, low quality, and low prices.

Meanwhile, in addition to ubiquitous and vocational education, we should encourage differentiated education based on individual student aptitude, giving outstanding students the opportunity to develop at a faster pace. There is a distinct disparity between education in China and the United States: In its transformation from an agricultural society to an industrial one, China advocates the training of engineers while the United States focuses on cultivating leaders. As we enter an era of digital and smart transformation, there is a huge demand for world-class strategists in the fields of politics, business, and the military as well as world-class leaders in technology and culture. These people should value mankind and human judgment and they should also value their country.

Education should revolve around teachers. Only outstanding teachers will be able to cultivate even more outstanding talent. In the past 30 years, government and society have made a massive investment in improving school equipment and facilities, but the welfare of teachers— especially those from the western and mid-western regions of China— has seen little improvement. Compared with knowledge workers in other fields, teachers' income is generally rather low. If teachers in the western and mid-western regions bring in as much money as those in Beijing and Shanghai, hordes of talent would flock into the northwestern and southwestern regions of China to teach. When teachers, our engineers of the soul, begin to enjoy the most prestigious status in society, there will certainly be more hope for China to make its second leap of development.

Reform dividends over the last three decades have been dividends of policy and population, but the success of future reform will depend on fully unleashing legal dividends and market dividends as well as equally important educational dividends. From this perspective, compared with developed countries in the West, China still has enormous development potential. Perhaps destiny will one day favor China once more.

In 1990, Alvin Toffler published *Powershift*, the third book in his trilogy. He argues that the United States will continue to have enormous advantages in the foreseeable future, but its position of absolute monopoly will be challenged and weakened. China will have more influence in the global economy and technology sectors. While following the rules and systems that the United States and Europe have established for economics and technology, China will play an important role in restructuring "the club" with new rules and new systems. Although there won't be any great shifts of power, power distribution and power sharing will be the new general trend.

Tian Tao is a member of the Huawei International Advisory Council and codirector of Ruihua Innovative Research Institute at Zhejiang University, Hangzhou, China. In 1991, Tao founded *Top Capital*, the first Chinese magazine on private equity investment, and has served as its editor-in-chief since then. He worked as a publisher of *Popular Science* (Chinese edition) from 1995 to 1997. In the past 20 years, Tao has started a number of businesses in advertising, publishing, and the media industry. He also cofounded two IT companies, Beijing Umessage Information Technology Co. and Hillstone Networks Co. Tao is a lover of several sports, including swimming and golf.

David De Cremer is the KPMG professor in management studies at Judge Business School, University of Cambridge, and an advisor to the Novartis AG ethics-based compliance initiative (Switzerland). He was named the most influential economist in the Netherlands (2009–10) and named one of the Global Top Thought Leaders in Trust (2016) by the organization Trust Across America. He has been awarded many international scientific awards, published more than 200 academic articles, and written popular books with his latest one being *The Proactive Leader: How to Overcome Procrastination and Be a Bold Decision-Maker*.

Wu Chunbo is a professor and PhD supervisor at Renmin University of China, Beijing. He earned his PhD in economics from the same university in 1998 and has served as the dean of the Institute of Organization and Human Resources, School of Public Administration, Renmin University of China. He has been serving as a senior corporate management advisor for Huawei since 1995. Playing tennis is his favorite hobby.

About the Authors